The Green Light

ALSO AVAILABLE FROM BLOOMSBURY

Ecosophical Aesthetics, edited by Colin Gardner and Patricia MacCormack
Eco-Aesthetics, Malcolm Miles
Ecosophy, Félix Guattari
The Three Ecologies, Félix Guattari
Lines of Flight, Félix Guattari
General Ecology, edited by Erich Hörl and James Burton
Environmental Ethics, Marion Hourdequin
The Ethics of Climate Change, James Garvey

The Green Light

A Self-Critique of the Ecological Movement

BERNARD CHARBONNEAU

Translated by
CHRISTIAN ROY

BLOOMSBURY ACADEMIC
LONDON • NEW YORK • OXFORD • NEW DELHI • SYDNEY

BLOOMSBURY ACADEMIC
Bloomsbury Publishing Plc
50 Bedford Square, London, WC1B 3DP, UK

BLOOMSBURY, BLOOMSBURY ACADEMIC and the Diana logo are trademarks
of Bloomsbury Publishing Plc

Original French text 'Le Feu vert: Autocritique du mouvement écologique'

© Bernard Charbonneau 1980

Copyright to the English translation © Christian Roy, 2018

Cover Design by Toby Way
Cover Image © Gustoimages / Science Source

All rights reserved. No part of this publication may be reproduced or
transmitted in any form or by any means, electronic or mechanical,
including photocopying, recording, or any information storage or
retrieval system, without prior permission in writing from the publishers.

Bloomsbury Publishing Plc does not have any control over, or responsibility for,
any third-party websites referred to or in this book. All internet addresses given
in this book were correct at the time of going to press. The author and publisher
regret any inconvenience caused if addresses have changed or sites have ceased
to exist, but can accept no responsibility for any such changes.

A catalogue record for this book is available from the British Library.

A catalog record for this book is available from the Library of Congress.

ISBN: HB: 978-1-3500-2708-4
PB: 978-1-3500-2709-1
ePDF: 978-1-3500-2707-7
ePub: 978-1-3500-2710-7

Typeset by Newgen KnowledgeWorks Pvt. Ltd., Chennai, India
Printed and bound in Great Britain

To find out more about our authors and books visit www.bloomsbury.com
and sign up for our newsletters.

CONTENTS

Foreword by Piers H. G. Stephens ix
Introduction by Daniel Cérézuelle xxi
Preface: The heart of the subject xxxi

PART ONE Seeds: The origins of the ecological revolt 1

1 Origins 3
   Prehistory of the ecological movement 3
   A great silence 8

2 Ecology, Year One 15
   Where the thesis, that is, science and America, generates its own antithesis 16
   The Green Light in France 22

3 The various constellations of the ecological nebula 29
   At the centre and on the margins of the maelstrom 29
   Where Nature switches from Right to Left 34
   Communal microcosms and silent majority 38

PART TWO Roots: Foundations of the ecological movement 43

4 Nature or freedom? 45
Nature 46
Freedom 48
The contradiction between nature and freedom 50

5 Nature and freedom 55
Nature and man united in the human environment: Town and country 55
Nature and freedom united in the existence of each human being 57
Nature and freedom associated in history 60

6 Nature and Christianity 65
The rupture of Creation 65
Incarnation 69
Christianity and the ecological movement 72

PART THREE Diseases and poisons: Contradictions and shortcomings of the ecological nebula 77

7 Nature, freedom and the ecological movement 79
The temptation of naturist fundamentalism 79
A critique of ecologism 84
The libertarian temptation 86
A critique of the ecological movement's anarchistic and non-violent strand 88
Beyond the ecological right and left 92

8 A fruit still green 95

Ecology without a doctrine 95
Weak points of ecological thinking 97
Shortcomings in economic and especially social reflection 105

9 Recycling 113

Creation or recyclable social by-product? 113
Recycling through fashion and fashionistas 115
Recycling through technostructure 118
Recycling through spectacle 122

10 Recycling through politicization–depoliticization 125

From political commitment to withdrawal from politics 126
Recycling through depoliticization 129
Recycling through politicization 133

PART FOUR Fruits: Sketch of an ecological politics 139

11 Topical utopia 141

Ecological conversion 141
A mediation between opposites 146
A revolution for that which exists 148

12 The ecological community 155

The personal basis 155
Language and ecological reason 159
Ecological society and meetings 161
The ecological order 167

13 Ecological politics 171

    Ecology and power 171
    Elements of ecological tactics in politics 176
    A non-economist economic policy 179
    Self-management and ecological self-sufficiency 183
    Agricultural policy and the dis-organization of leisure 186

Envoi 193

Index 195

FOREWORD

Piers H. G. Stephens

> Nature is not a luxury, as some would have us believe;
> it is like water, air, freedom, which man cannot
> do without or else he dies.
>
> – BERNARD CHARBONNEAU

At a time when some anglophone environmental thinkers are declaring that nature is dead or ended and that this is nothing to regret, or that an incoming Anthropocene era should mandate an extended human managerialism, or even that the ideals and canon of traditional Western environmentalism were fundamentally misconceived,[1] it is both a privilege and a pleasure to introduce the translated work of a thinker who anticipated and challengingly interpreted many of the social and ecological dynamics with which environmentalism is now struggling. As Daniel Cérézuelle eloquently explains in his foreword, Bernard Charbonneau's early insights into the dynamics of modern society began to be publicly formulated in the 1930s, and in alliance with Jacques Ellul, he identified from the first the central importance of science and technology in establishing an ever-accelerating pattern of socioecological transformation, a set of dynamics that play out in similar ways under all of the

---

[1] Steven Vogel, *Thinking Like a Mall* (Cambridge, MA: MIT Press, 2015); Earle Ellis, 'Neither Good nor Bad', *New York Times*, 2011. https://www.nytimes.com/roomfordebate/2011/05/19/the-age-of-anthropocene-should-we-worry/neither-good-nor-bad; Paul Kingsnorth, 'Rise of the Neo-greens', in George Wuerthner, Eileen Crist and Tom Butler (Eds), *Keeping the Wild: Against the Domestication of Earth* (Washington, DC: Island Press, 2014), 3–9.

mainstream political ideologies. In this late work, *The Green Light*, first published in French in 1980 and beautifully translated here by Christian Roy, we see the remarkable prescience with which Charbonneau was able to anticipate some of the oncoming problems of modern society and of environmentalism's challenge to it. Despite the numerous significant changes that have occurred in the intervening time, most notably the collapse of communism and the rise of deregulatory market right-wing politics, Charbonneau's diagnosis remains strikingly fresh and apposite to our times – a point that in itself suggests that the core of his social critique, with its emphasis on technoscientific influences, may well have a high degree of accuracy. Accordingly, let us look at some of his key ideas in relation to current debates within environmental philosophy and green political theory.

Technology, nature and society

The most obvious starting point here, expounded in more detail in Daniel Cérézuelle's foreword, is Charbonneau's concept of the Great Moulting, a stage towards which the expansionist dynamics of technological transformation, yoked to states and markets, have been transforming society since the advent of industrialization. This stage is one in which humanity is shedding nature as its environment as a snake sheds its skin, and Charbonneau strikingly states that it 'is developing on its own, out of the collective mind' and that it is more 'than a social or even a biological fact, it is a geological one' (*The Green Light*, p.13). This vision clearly anticipates on a larger scale the critiques and explanations of ongoing ecological destruction put forward in more recent years by figures like Albert Borgmann and Anthony Weston, with their respective accounts of the 'device paradigm' and the embedded tendencies towards self-validating reduction in our contemporary dealings with nature.[2] But more than this, the similarity here to recent claims of our having arrived in an Anthropocene era is manifest. Charbonneau's response

---

[2] Albert Borgmann, *Technology and the Character of Contemporary Life* (Chicago: University of Chicago Press, 1984); Anthony Weston, 'Self-Validating Reduction: Toward a Theory of Environmental Devaluation', *Environmental Ethics*, 18, 1996, 115–32.

to the prospect is valuable indeed, being radically different from the technocentric boosters who have jumped upon that concept, as well as more nuanced and wide-ranging than many of its critics. For Charbonneau, the objection to such a prospect does not rest upon appeals to wilderness or the intrinsic value of nature. Rather, he intimately ties the process of the Great Moulting, and thus indirectly the value of nature, to the reduction of human freedom, for the ongoing dynamic of the Moulting threatens nature and freedom together. The technoscientific drives to socially organize, centralize and economically transform nature necessarily mean more specialization of knowledge and labour, more regulation, greater separation of human beings from the natural sources of their physical, emotional and spiritual sustenance. As a founder of Gascon personalism, Charbonneau's thought stands at the most strongly individualist, anti-authoritarian and politically regionalist point among the various forms of personalism that emerged in 1930s France.[3] Both he and Ellul were struck, as personalists, by the ways in which mass industrial society can construct people as members of a crowd rather than a reflective community, how it exposes them to manipulation and, in Charbonneau's case, how it frustrates necessary modes of expressing human freedom. Conceptually recognizing the deep affinity between nature and spontaneity, Charbonneau's analysis thus critically focuses on the technological reduction of the very ideal that proponents of our contemporary political order often profess to champion: individual liberty.

I suggest that part of the attraction of this perspective today could be that it retains most of what is radical in traditional green thought while bypassing the tangled philosophical debates over intrinsic value, and the increasingly tired intellectual clashes between anthropocentrism and nonanthropocentrism. Though Charbonneau recognizes humans as social beings, he also maintains that they 'need to experience the otherness of nature in order to exert their freedom in a personal way (which is not necessarily a solitary activity), and they need also to experience sensuously both

---

[3]For an overview in English of the distinctive place of Charbonneau and Ellul among the various strands of the French personalist movement, see Christian Roy, 'Ecological Personalism: The Bordeaux School of Bernard Charbonneau and Jacques Ellul', *Ethical Perspectives*, 6, no. 1, April 1999, 33–44. http://www.ethical-perspectives.be/page.php?LAN=E&FILE=ep_detail&ID=21&TID=1363.

their own freedom and the richness of the world'.[4] Nature is thus both a wider frame in which humans need to act in order to manifest their freedom, and a touchstone of that freedom when set against the dynamics of technological modernity. These observations clearly anticipate significant reflections in recent anglophone environmental philosophy on the tensions between our affinity for nature and the worth of its difference,[5] and Charbonneau insists that there are positive ways in which we can and should properly recognize nature's otherness. But this is not a simple matter, and he is acute in rejecting easy assumptions that either collapse humans into nature or place us into a simplistic dualism: 'Man is not nature *or* supernature, nature *or* culture, but both the one *and* the other. And each time man allows himself to be reduced to just one of these terms, he negates himself and turns to stone, so to speak' (p.53). In a certain sense, humanity first moves outside nature by means of attaining a second nature, society, and there is thus an essential tension between our first and second natures, a notion which goes back to Rousseau but one to which Charbonneau gives eloquent refinements and ecological updating. In doing so, he gives his own distinctive twists to both normative and ontological issues, offering fresh perspectives on our current debates in this area.

Nature, freedom, development

Much of what is most valuable in the core of *The Green Light*, in my view, comes in Charbonneau's intricate reflections on the relationship between nature and freedom in Chapters 4 and 5. Long before the extended debates over social constructionism

---

[4] Daniel Cérézuelle, 'Nature and Freedom: An Introduction to the Environmental Thought of Bernard Charbonneau', in Bruce V. Foltz and Robert Frodeman (Eds), *Rethinking Nature: Essays in Environmental Philosophy* (Bloomington: Indiana University Press, 2004), 322.
[5] See, for example, Robert E. Goodin, *Green Political Theory*, Cambridge, UK: Polity Press, 1992; Neil Evernden, *The Natural Alien: Humankind and Environment* (Toronto: University of Toronto Press, 1993); Andrew Biro, *Denaturalizing Ecological Politics: Alienation from Nature from Rousseau to the Frankfurt School and Beyond* (Toronto: University of Toronto Press, 2005); Simon Hailwood, *Alienation and Nature in Environmental Philosophy* (Cambridge: Cambridge University Press, 2015).

within environmental thought that have animated the field in recent years, he notes the following paradox: our capacity to linguistically represent nature means that our idea of it 'becomes a product of culture', so that, to 'be logically consistent, an ecologist would have to remain mute like the plants'; yet he recognizes that the paradox must be embraced, for 'only man can defend nature against man' (p.59). A condition of being motivated to this defence is to validate individual experience within the realm of nature, especially the sense of appreciation that comes with solitude, and here Charbonneau regards the shift to collective or commercialized experience, such as tourism, to be disastrous for nature. Instead, he places himself on the side of those whose inspiration comes from personal feeling in nature, seeing this as experientially basic to the broader political challenge of opposing the instrumentalizing dynamics of technoscience, the state and contemporary political orthodoxies: 'Insofar as it remains personal, only this feeling can save nature ... from now on, he [man] can only respect nature out of love, and nothing can force him to take such a decision' (p.60). In this regard, Charbonneau's ecological-minded personalism has some whiff of tradition, for it recalls not only Rousseau and Thoreau but also a series of thinkers whom he leaves unmentioned, despite his occasional sympathies for the Anglo-Saxon individualist tradition: one thinks of the English Romantics Wordsworth and Coleridge, of John Stuart Mill, of Emerson and of William James, all of whom share in different ways Charbonneau's emphasis on the informing and transformative direct tang of individual nature experience, and on its importance for human fulfilment.

However, though there is also shared ground in the deep scepticism towards industrialization and the state shown by several of these thinkers, the distinctness of Charbonneau's vision is underlined by the fact that he would strongly reject the political liberalism with which much of this tradition of thought is associated. To use the terms popularized during the debate within the German Greens, Charbonneau is a Fundi rather than a Realo, one who is painfully aware of the extent to which conventional politics is embedded in a technoscientific dynamics and who is deeply suspicious of the danger of co-optation. For him, as for Ellul, the emergence of modern technology as a threat to the values of freedom and equality first emerged from the seedbed of liberalism's own belief in expanded economic productionism and its attempts to generate freedoms that transcend nature; on this view, political ecologists now face a situation in which all the opposing points on the conventional political spectrum share a fixation on quantifiable measurement, an unquestioned support for endless economic growth – 'development', in Charbonneau's terms here – and a set of overarching technoscientific dynamics that can only be reversed by a revolutionary alternative. (Indeed, Ellul even went so far as to paint 'Fascism as Liberalism's Child' in the personalist review *Esprit* in 1937 on precisely these lines.[6]) By this account, recent attempts to disconnect the liberal tradition from its economically expansionist and technologically optimistic roots will necessarily be doomed, as the operations of technological modernity have already acquired an expansionist logic of their own, one so pervasive that it is even embedded in the mechanisms of liberalism's communist and fascist opponents. Only by beginning with direct experience and personal connection to nature can opposition root itself, see beyond and challenge the technocratic framework: 'The active awareness of nature is a matter of morals, or rather of ethics. The protection of nature is rooted less in matter than in freedom' (p.86). Without a successful bottom-up regionalist challenge from political ecology, itself informed by direct experience and the important liberties of independence, spontaneity, local self-governance and autonomous self-development, nature and non-economic liberties will, in Charbonneau's view, always be sacrificed to the demands of technoscientific organization and bureaucracy. Though readers familiar with the recent debates about the relationship between liberalism and ecologism may know that I have been somewhat more hopeful about the resources of the liberal tradition than Charbonneau was,[7] anyone who has watched

---

[6]Jacques Ellul, 'Le fascisme, fils du libéralisme', *Esprit*, no. 53, February 1937, 761–97, reissued in *Cahiers Jacques Ellul, Pour une critique de la société technicienne*, no. 1: *Les années personnalistes*, 2004, 113–37. A translation of this key text of the Bordeaux School is set to open an upcoming book edited by David Gill and David Lovekin, *Political Illusion and Reality: Engaging the Prophetic Insights of Jacques Ellul* (Eugene, OR: Wipf and Stock, 2018). I am indebted to Christian Roy for drawing my attention to this article.

[7]Some brief key texts here include Marcel Wissenburg, *Green Liberalism: The Free and the Green Society* (London: UCL Press, 1998); John Barry and Marcel Wissenburg (Eds), *Sustaining Liberal Democracy: Ecological Challenges and Opportunities* (London: Palgrave Macmillan, 2001); Piers H. G. Stephens, 'Green Liberalisms: Nature, Agency and the Good', *Environmental Politics*, 10, no. 3,

in consternation the multiple failings of the contemporary political order to deal with the threats of climate change and biodiversity loss (to name but two crises) and the inability of environmental reformists to win meaningful victories cannot help but seriously consider that Charbonneau's pessimism may have been justified. If that is the case, then both Charbonneau's case for a grounding in personal experience and his insistence that the defence of nature necessitates rejecting all our existing technologically orientated ideologies deserve the closest of attention. For, in identifying the global character of the threats to nature and human freedom, and in insisting on radical bottom-up alternatives, he may be seen to have long predated Naomi Klein in perceiving a systemic totalizing threat and effectively declaring that 'this changes everything'.[8]

History and the ontology of nature

One possible objection to Charbonneau's thesis is to historically challenge the view that nature is of great importance to the qualitative dimensions of human life and freedom. It is, for instance, a common claim that nature appreciation is a modern phenomenon that emerged only as a side effect of affluence, as is asserted by Roderick Nash about the American wilderness tradition, for example.[9] But in broader terms this view is demonstrably false, and as a generalized observation about nature appreciation, especially in

---

2001, 1–22; Marius de Geus, *The End of Overconsumption: Towards a Lifestyle of Moderation and Self-Restraint* (Utrecht: International Books, 2003); Simon Hailwood, *How to Be a Green Liberal* (Chesham, UK: Acumen, 2004); John M. Meyer, 'Green Liberalism and Beyond', *Organization and Environment*, 18, no. 1, 2005, 116–20; Derek Bell, 'Liberal Environmental Citizenship', *Environmental Politics*, 14, no. 2, 2005, 179–94; John M. Meyer, 'We Have Never Been Liberal: The Environmentalist Turn to Liberalism and the Possibilities for Social Criticism', *Environmental Politics*, 20, no. 3, 2011, 356–73; Piers H. G. Stephens, 'Environmental Political Theory and the Liberal Tradition', in Teena Gabrielson, Cheryl Hall, John M. Meyer and David Schlosberg (Eds), *The Oxford Handbook of Environmental Political Theory* (Oxford: Oxford University Press, 2016), 57–71.
[8] Naomi Klein, *This Changes Everything: Capitalism vs the Climate* (New York: Simon & Schuster, 2014).
[9] 'Nature appreciation is a "full stomach" phenomenon that is confined to the rich, urban and sophisticated'. A quote from Roderick Frazier Nash, *Wilderness and the American Mind* (5th edition) (New Haven, CT: Yale University Press, 2014), 343.

terms of social critique, it misses much. Writing more than twenty years before the first edition of Nash's book, George Orwell noted the similar claim that nature appreciation was only possible because of industrialization, and so to 'love the country one must live in the town, merely taking an occasional weekend ramble at the warmer times of year'. But, like Charbonneau, and contra Nash and others, Orwell rejected this thesis, instead noting abundant countervailing evidence: medieval literature 'is full of an almost Georgian enthusiasm for Nature, and the art of agricultural people such as the Chinese and Japanese centers always round trees, birds, flowers, rivers, mountains'.[10] Further examples are not hard to find. In music, for instance, the oldest surviving example of independent melodic counterpoint is likely to be the medieval English rota *Sumer Is Icumen In*, a celebration of nature's zest and beauty that dates back to the later thirteenth century.[11] Far from being a solely modern product, the presence of engagement with nature as a source of joy, vitality, liberty and psychological balance was widespread and provocative in premodern thought, and part of Charbonneau's historical thesis deals precisely with the way in which this affirmation and regular presence was transformed through commercial industrialization, a process of 'forcing back into luxury ... and leisure reserved for the rich what was the expression of a vital need for air and freedom', thus making nature 'assimilated to its opposite: to refinements of literature and Art, to the entertainments of princes ... and then of the bourgeois' (p.5). It was against this backdrop that the processes of industrial technique were to emerge, and the steps to the Great Moulting began to be taken.

Here too one can see insightful moves from Charbonneau that are well worth attention for contemporary debates. Though he is reluctant to discuss ontology in conventional terms, there is clear evidence of the points at which the techniques of artifacticity cross the boundary, of how a certain sort of instrumental reason ontologically distinguishes its products through its transformations: 'Nature? You discover it the day you come out of it: it was distinguished from culture when the sky that was once

---

[10] George Orwell, 'Some Thoughts on the Common Toad', in George Orwell, *Orwell and Politics* (ed. Peter Davison) (London: Penguin, 2001), 447–8.
[11] Bella Millett, 'Sumer Is Icumen In': London, British Library, MS Harley 978, f. 11v: The Manuscript, 2004.

held up by the world ash tree Yggdrasil became enclosed within the sphere of steel, bricks and smoke of Krupp's Germany. Even if we do not know what nature is, we know what it is not: the discarnate order of human laws imposed by the iron rule of barracks and police stations, the sharp-toothed machinery that tears down oak trees and tears open hillsides, the grey concrete glacier that covers them, the stinking cloud that rots the air and water. In a word, human artifice' (p.46). But at the same time, Charbonneau is on guard against the dualisms embedded in an association of nature only with untouched pure wilderness, seeing that such an equation, consistently applied, would be no threat to the ongoing dynamics, but would instead fit easily within a technologically managerialist ethos. Rather, he invokes aspects of the agrarian tradition, more familiarly represented in different ways by peasant culture in France and Wendell Berry in the United States,[12] a tradition that acknowledges the close and vital bond of humanity to the land even as it repudiates the excesses of human manipulation. 'I can yield to the human urge, at once legitimate and usurped, to create a work and to fashion the soil as I wish; this is also called exerting my power, a passion that can degenerate into madness', he observes, but critically anticipating those who would today wish to drop the very idea of nature from environmental thought, he refuses any such denial. 'Nature is everywhere ... around me', Charbonneau insists, whereas 'I dare not say my "environment", since this fashionable term suggests that we do not really know what it is and that it is exterior to me' (p.47). His emphasis on the vagueness of 'environment' here tacitly points to the enormity of possible manipulations that a green politics stripped of the notion of nature could allow, and suggests that Charbonneau would rather sympathize with those thinkers who look at ontological naturalness and artifacticity along a spectrum, with historical and psychological factors incorporated.[13] Certainly, as genetic modification, climate change and other deep technologies

---

[12]See Daniel Cérézuelle, 'Wendell Berry et Bernard Charbonneau, critiques de l'industrialisation de l'agriculture', in *Encyclopédie de l'Agora* (29 January 2003). http://agora.qc.ca/encyclopedie/index.nsf/Impression/Agriculture–Wendell_Berry_ et_Bernard_Charbonneau_critiques_de_lindustrialisation_de_lagriculture_par_ Daniel_Cerezuelle.

[13]See, for example, Keekok Lee, *The Natural and the Artefactual: The Implications of Deep Science and Deep Technology for Environmental Philosophy* (Lanham, MD: Lexington Books, 1999); Piers H. G. Stephens, 'Nature, Purity, Ontology',

and technological impacts increasingly transform the world into a wall of mirrors reflecting human manipulations, it is hard to envision Charbonneau not fervently opposing the reduction, and equally hard to deny his perspicacity in foreseeing, far in advance, many of the dynamics involved.

There are plenty of other points that could be made about Charbonneau's visionary analysis as it unfolds in this remarkable and beautifully crafted book. A close reading will reveal that there is much more still that he anticipates, including the tensions between the direct and local character of much environmental motivation and the need for reining in ecologically destructive dynamics at the global level, and ranging from the emergence of a professional political class to the importance of food ethics and localism, from the decontextualizing tendencies of modern technology to reflections on the meaning of human-centredness. But premature clamour should not be allowed to dull a fine edge: read Charbonneau for yourself in this fine translation by Christian Roy. I believe anglophone green thinkers will soon find themselves very glad to at last have the opportunity.

References

Barry, John and Marcel Wissenburg (Eds), *Sustaining Liberal Democracy: Ecological Challenges and Opportunities*, London: Palgrave Macmillan, 2001.

Bell, Derek, 'Liberal Environmental Citizenship', *Environmental Politics*, 14, no. 2, 2005, 179–94.

Biro, Andrew, *Denaturalizing Ecological Politics: Alienation from Nature from Rousseau to the Frankfurt School and Beyond*, Toronto: University of Toronto Press, 2005.

Borgmann, Albert, *Technology and the Character of Contemporary Life*, Chicago: University of Chicago Press, 1984.

Cérézuelle, Daniel, 'Wendell Berry et Bernard Charbonneau, critiques de l'industrialisation de l'agriculture', in *Encyclopédie de l'Agora*, 29 January 2003. http://agora.qc.ca/encyclopedie/index.nsf/Impression/ Agriculture–Wendell_Berry_et_Bernard_Charbonneau_critiques_de_ lindustrialisation_de_lagriculture_par_Daniel_Cerezuelle.

Environmental Values, 9, no. 3, 2000, 267–94; Helena Siipi, 'Dimensions of Naturalness', *Ethics and the Environment*, 13, no. 1, 2008, 71–103.

Cérézuelle, Daniel, 'Nature and Freedom: An Introduction to the Environmental Thought of Bernard Charbonneau', in Bruce V. Foltz and Robert Frodeman (Eds), *Rethinking Nature: Essays in Environmental Philosophy*, Bloomington: Indiana University Press, 2004.

Ellis, Earle, 'Neither Good nor Bad', *New York Times*, 2011. https://www.nytimes.com/roomfordebate/2011/05/19/the-age-of-anthropocene-should-we-worry/neither-good-nor-bad.

Ellul, Jacques, 'Le fascisme, fils du libéralisme', *Esprit*, no. 53, February 1937, 761–97, reissued in *Cahiers Jacques Ellul, Pour une critique de la société technicienne*, no. 1: *Les années personnalistes*, 2004, 113–37.

Evernden, Neil, *The Natural Alien: Humankind and Environment*, Toronto: University of Toronto Press, 1993.

de Geus, Marius, *The End of Overconsumption: Towards a Lifestyle of Moderation and Self-Restraint*, Utrecht: International Books, 2003.

Goodin, Robert E., *Green Political Theory*, Cambridge, UK: Polity Press, 1992.

Hailwood, Simon, *How to Be a Green Liberal*, Chesham, UK: Acumen, 2004.

Hailwood, Simon, *Alienation and Nature in Environmental Philosophy*, Cambridge: Cambridge University Press, 2015.

Kingsnorth, Paul, 'Rise of the Neo-greens', in George Wuerthner, Eileen Crist and Tom Butler (Eds), *Keeping the Wild: Against the Domestication of Earth*, Washington, DC: Island Press, 2014, 3–9.

Klein, Naomi, *This Changes Everything: Capitalism vs the Climate*, New York: Simon & Schuster, 2014.

Lee, Keekok, *The Natural and the Artefactual: The Implications of Deep Science and Deep Technology for Environmental Philosophy*, Lanham, MD: Lexington Books, 1999.

Meyer, John M., 'Green Liberalism and Beyond', *Organization and Environment*, 18, no. 1, 2005, 116–20.

Meyer, John M., 'We Have Never Been Liberal: The Environmentalist Turn to Liberalism and the Possibilities for Social Criticism', *Environmental Politics*, 20, no. 3, 2011, 356–73.

Millett, Bella, 'Sumer Is Icumen In': London, British Library, MS Harley 978, f. 11v: The Manuscript, 2004.

Nash, Roderick Frazier, *Wilderness and the American Mind* (5th edition), New Haven, CT: Yale University Press, 2014.

Orwell, George, 'Some Thoughts on the Common Toad', in George Orwell, *Orwell and Politics*, ed. Peter Davison, London: Penguin, 2001.

Roy, Christian, 'Ecological Personalism: The Bordeaux School of Bernard Charbonneau and Jacques Ellul', *Ethical Perspectives*, vol.

6, no. 1, April 1999, 33–44. http://www.ethical-perspectives.be/page.php?LAN=E&FILE=ep_detail&ID=21&TID=1363.

Siipi, Helena, 'Dimensions of Naturalness', *Ethics and the Environment*, 13, no. 1, 2008, 71–103.

Stephens, Piers H. G., 'Nature, Purity, Ontology', *Environmental Values*, 9, no. 3, 2000, 267–94.

Stephens, Piers H. G., 'Green Liberalisms: Nature, Agency and the Good', *Environmental Politics*, 10, no. 3, 2001, 1–22.

Stephens, Piers H. G., 'Environmental Political Theory and the Liberal Tradition', in Teena Gabrielson, Cheryl Hall, John M. Meyer and David Schlosberg (Eds), *The Oxford Handbook of Environmental Political Theory*, Oxford: Oxford University Press, 2016, 57–71.

Vogel, Steven, *Thinking Like a Mall*, Cambridge, MA: MIT Press, 2015.

Weston, Anthony, 'Self-Validating Reduction: Toward a Theory of Environmental Devaluation', *Environmental Ethics*, 18: 1996, 115–32.

Wissenburg, Marcel, *Green Liberalism: The Free and the Green Society*, London: UCL Press, 1998.

INTRODUCTION

Daniel Cérézuelle

Translated by Christian Roy

In 1937, Bernard Charbonneau published in an obscure French personalist newsletter an article entitled 'Le sentiment de la nature, force révolutionnaire' ('Feeling for Nature as a Revolutionary Force'); today this essay is considered the first French text of the environmental movement, or what is called in France political ecology. As early as the 1930s, Charbonneau developed an original philosophy of freedom and nature, and he is now considered by intellectual historians to be one of the founders of the French ecological movement. Charbonneau, who was born in 1910 and died in 1996, was two years older than Jacques Ellul (1912–1994), his lifelong friend. Together, they founded what some historians of ideas term 'Gascon personalism'. Indeed, Ellul used to say that Charbonneau taught him to 'think and to be a free man'. However, the truth is that even in France Charbonneau remains little known, although there is growing interest in his work.

The central intuition

All of Charbonneau's works are inspired by the central conviction that modern technoscientific development is bringing about what he calls *la Grande Mue*: literally, 'the Great Moulting'. His concern with science and technology is related to the fact that, being born in 1910, he grew up under the shadow of the 'Great War' (1914–1918), the first truly industrial war, with its millions of casualties. He was soon convinced that since that war, humankind was experiencing an utterly new phase of its history: more than a sea change, a change of scale so vast that it fundamentally altered it, causing the species to shed its old Neolithic skin in what amounted to a Great Moulting, with two main characteristics.

First, the Great War as a total war introduced a complete subordination of reality to the logic of technological and industrial imperatives. This logic requires the mobilization of the whole population, of all resources (industry, agriculture, forests) and therefore of all space. The war also achieved a mobilization of the inner life of the people. Whichever the side, not only were vast populations affected by the war, but people were also drawn into collective consent to it as a matter of course, thus justifying the anonymous process that was destroying them as individuals. The Great War was the first experience of what Charbonneau terms a 'total social phenomenon'. He insists that such a phenomenon does not have to be totalitarian in order to be total. Under such circumstances, in contrast to humanistic assumptions about individuality, the so-called inner self spontaneously conforms to anonymous social trends. By nature, social facts transcend individual human consciousness, and freedom is difficult to achieve against what Charbonneau calls the 'order of things' (*l'ordre des choses*), including, in the first place, the social nature of human beings themselves.

Second, this 'Great Moulting' is characterized by self-acceleration. Since the Great War, human power has been taking hold of the entire planet at an ever-increasing pace. Charbonneau argues that this acceleration is a quasi-autonomous process: it is not directed by a collective project, most of its effects have not been chosen and there is no pilot; it simply rushes forward. Change happens, period. This mutation cannot be compared with previous major social transformations, such as the reformations brought about by revolutions in which collective behaviour was reorganized by new meanings and values. It is more like a biological mutation or, rather, like a geological phenomenon such as a huge landslide: powerful, all-encompassing, controlled neither by thought nor by any human intention.

Beginning in August 1914, humankind crossed a threshold, entered a new phase of history. From a young age, since the late 1920s (I obtained this information from Ellul, who met him while they were attending *lycée* in Bordeaux), Charbonneau was convinced that contemporary conflicting ideologies (socialism, fascism, communism, nationalism, liberalism, anti-colonialism, etc.) were already outdated and provided no purchase on this Great Moulting, which was occurring apparently without the awareness of his fellow citizens. From then on, he spent all his intellectual energy to help his contemporaries realize that this uncontrolled development of industry, technology and science was the *problem* and not the solution or a neutral means for various social and political projects. He became very active in French intellectual circles, especially among those who today are called the 'non-conformists of the 1930s'.

Towards a political ecology

Instead of just aiming for a political reorganization of society, these groups were developing a critique in terms of civilization, of the material as well as the spiritual structures of society. Participating in this discussion gave Charbonneau and Ellul the opportunity to clarify what was essential to them: the conflict between a goal of personal freedom and the Great Moulting. This has been the specific orientation of their Gascon personalism, and a task that both of them would pursue for the rest of their lives. It is within this framework that Charbonneau would come to see the issue of nature as crucial.

As it turned out, Charbonneau failed to create a new political movement because at that time the whole Western world was obsessed with ideological conflicts. No one took seriously his critical analysis of the limits of techno-industrial development. It was only in the 1970s that the fledgling French environmental movement discovered that there was someone who had spent much time thinking about the issues that it was raising. This is one of the reasons why most of Charbonneau's books, which had been written during the 1940s and 1950s, were published much later. For example, his book *L'État* ('The State'), written in 1947, was not published until 1987. His major book on freedom, *Je fus* written in the 1940s, had to wait until 1980, and even then, was only privately published. Charbonneau's thought about the relationship between nature and freedom in modern times is developed in the following six works: *Le Jardin de Babylone* (1969), *Tristes campagnes* (1973),

Le Système et le Chaos (1973), *Le Feu vert* (1980), *Je fus* (1980) and *Finis Terrae* (posthumous, 2010).

Industrial dehumanization

'Feeling for Nature as a Revolutionary Force' was written by Charbonneau in 1937 and circulated as a fifty-three-page manifesto by the Gascon personalist group. At that time, Charbonneau was twenty-seven years old and his text bears some of the imperfections of an early work; nevertheless, it also powerfully states some important ideas. First of all, it proclaims that this attraction for nature is the expression of a revolt that is more radical than modern political programs. Fascism, communism and liberalism are superficial answers to the problems of industrial civilization; they will not change our daily life, the main reason being that 'they all share the same religion of production and industrial organization'.

Charbonneau argues that the modern obsession with production and economic power sooner or later comes into conflict with our need for freedom, because economic and industrial progress can be obtained only by a refinement of the *armature sociale* or 'societal framework'. Therefore, 'the synthesis between endless progress in freedom and an endless growing comfort is a utopia'. Collective impersonal disciplines, at the cost of personal freedom and responsibility, are the price that must be paid for economic progress.

In order to liberate themselves from nature, since their origins, human beings have created tools and weapons. The most effective tool is association: society is a necessity imposed on individuals by nature itself. Human beings are therefore social – or, to speak like the ancients, political animals. But collective power over nature has a price: it is paid by political and economic power over the individual. The development of Western science and technology has exacerbated this situation: on the one hand, a greater efficacy has been achieved in the exploitation of nature; on the other, this efficacy is obtained at the price of a greater control over human beings: increased division of labour and the concentration and complication of human activities.

All of Charbonneau's books remind us that today this trend is 'as inflexible as the course of the stars'. Everywhere, technological and economic development results in the growth of bureaucracy and of its impersonal and abstract organization. This trend has nothing to do with ideologies, political philosophies and constitutions. Countries that wish to limit public bureaucracies controlled by the state must rely on private bureaucracies and the proliferation of files, databases, procedural paperwork managers, executives and secretaries. All our cities are full of offices: we build towers for them, more and more. Of course, this universal state of affairs inflicts violence on the individual, because bureaucracy, in its essence, cannot know the individual. Sometimes we react against such an impersonal violence. But when we protest, we regularly do so in the name of increased efficiency and we encourage thereby a further development of bureaucracy – of a new bureau, with its technicians, computers, better forms and paperwork and, thus, another step towards the depersonalization of life.

System and chaos

Techno-economic development and industrial organization foster the bureaucratization of life, but this is only one side of the modern predicament. Rapid development also has a disorganizing influence and, as a result, all societies are exposed to a dangerous dialectic between system and chaos.

Since human knowledge is limited, since our goals and behaviours are far from being reasonable – let alone rational – the growth of our technical power on nature and society necessarily results in disorganizing effects both in nature and in society. A historian and geographer by training, very early on Charbonneau devoted much energy to identifying and analysing the multidimensional costs of progress and their chaotic results: destruction of the landscape by industrial farming; destruction of local societies condemned to disappear by the logic of techno-economic development; cultural crisis. The constant invention of new technologies and the acceleration of technological change force us to live in a world of accelerated obsolescence, which renders traditional ways of life ineffective. Indeed, this permanent process of change takes place so rapidly that it deprives us of the time necessary for creating new ways of living together that might control our technology and put our values into practice.

Charbonneau tells us that the costs of the technologization of life and economic development tend to foster a further acceleration of change and a reinforcement of controls. Techno-economic development results in organizations that are more and more powerful, but also more and more complex and, therefore, fragile. In such a world, we are forced to intensify social organization in order to manage diminishing resources such as clean air, water, forests and, more generally, diminishing space. It is also necessary in order to anticipate and control human reactions as an obstacle to such management, which must operate on an increasingly global scale.

But the globalization of techno-economic development and control stimulates micro-nationalisms, which contradict large-scale rational management. Such a situation calls for new worldwide as well as local controls. Charbonneau warns that since we are building a world of powerful, complex and vulnerable technologies, we are obliged to anticipate everything – not only about nature, but also about human beings. Hence, the development of the social sciences and the proliferation of new kinds of experts and specialists for new bureaucracies. And since techno-economic development has worldwide consequences, only a worldwide organization (something like a world state) capable of managing the whole planet can avoid ecological and political chaos.

Unfortunately (or fortunately?), it is not clear that such an organization can be achieved, precisely because of the rise of national, religious, cultural and other regional forces as stimulated and exacerbated by globalization. But facing disorganization and the risks of chaos, we know only one response: more science, more technology, more experts for organizing social life. If this reinforcement of the systematic and impersonal character of social life succeeds, we will be able to pursue development a few steps further; new innovations, new changes will foster new risks and more frustrations, which will call again for more organization and social control, to which individuals and groups will respond with more radical claims.

We are therefore facing a paradoxical situation: technoscientific progress and economic development were initiated by a spirit of freedom, but they result in a tendency towards strict social organization. The need for personal freedom, initiative and responsibility may be the great loser in the blind process to which it has given rise. Either ecological and social chaos or scientific and bureaucratic management of society: none of these outcomes is acceptable to those attached to their land and to their freedom.

Nature as a condition of freedom and the ambiguities of ecology

In a world that tends to become totally organized by impersonal processes, the protection of nature is a vital necessity – not only so as to avoid ecological disasters, but also to preserve freedom. Charbonneau's ecological thinking is distinctive in that he reminds environmentalism of its duty to act in view of two values: maintaining a safe ecological order and responding to the human need for personal freedom. Achieving a balance between those two different – though equally necessary – goals is certainly a difficult task, a somehow 'unnatural' endeavour, but it is essential for the future of humankind, because the human condition is ambiguous. Tension is inherent to the human condition, and freedom is the achievement of a precarious balance between opposite requirements.

Charbonneau has been very intent to warn environmentalists about some inner ambiguities that may result in a betrayal of both nature and freedom. Motivated initially by a reaction against industrial and social organization, environmentalism may end up legitimizing the perfecting of even more rigorous and encompassing organization.

This attitude may lead to the many difficulties that Charbonneau analyses in *Le Feu vert* (1980). Let me briefly mention only one, which has already come up: an exclusive emphasis on the protection of 'pure nature' or wilderness. Charbonneau warns that such a concern for a nature totally distinct from culture suits perfectly our social organization, which excels in the division of functions and in zoning. In that case, environmentalism will end up merely giving the 'green light' for the development of a new specialized branch of our industrial and technological world. As he writes,

> The idea of a nature perfectly distinct from culture suits a society that practices the division of functions and zoning. Ecologism is the most fitting ideology for the handful of scientists and civil

servants in charge of managing the tiny sector of a chemically pure nature from which man – but for the licensed naturalist – is excluded.

The nature fundamentalist can be seamlessly integrated into the industrial system as a manager of nature reserves or of national parks (the regional Luna Park hardly being more than a bastard of green space and trade fair) that serve as an alibi for industrial, real estate, land or tourism reserves, at a ratio of one lark for one horse. (p.83)

From these analyses, Charbonneau draws the conclusion that if we are to avoid chaos or total organization, and if we take seriously our need for freedom and nature, we must break with the logic of accelerated development. Such a break would certainly require the reorientation of our civilization, which Charbonneau was already advocating in the 1930s. Nature and freedom: two serious reasons for setting limits to scientific, technological and industrial development. But since no science can tell where the limits should be set – because specialized science does not know about freedom, only about necessity, the search for these limits must be everyone's concern.

Translating Bernard Charbonneau

Charbonneau's books are carefully crafted. Convinced that common language is the most appropriate medium for addressing issues which should be everyone's concern, he avoids technical terms. But his style is sometimes disconcerting because of its condensation, inherited from his classical schooling. He is often ironic or poetic. Furthermore, Charbonneau relies on a vast culture which is not always shared by his readers. Today, how many of them know about the cosmic tree Yggdrasil mentioned in Chapter 4 of *Le Feu vert*? It is true that Charbonneau is not easy reading and requires keen attention. But here is the statement of Lani Niles, a professional reader who, in 1985, was asked by Harper & Row to make an assessment of Charbonneau's most philosophical book: *Je fus. Essai sur la liberté* ('I was. An Essay on Freedom').

Just a brief statement of my own evaluation: I expected to be thoroughly bored by the book, since the chapter headings sounded pretentious, but I found the ideas to be significant, original, and provocative for my own life. The book is developed very logically and I found most of the concepts to be clearly and simply explained, though the subject itself is not easy and the book would not be accessible to everyone. I can't speak to its market value, but I can say that I think that the book is important and more than merits publishing. Thank you for the opportunity to read it![1]

Translating Charbonneau's beautiful and subtle French into English requires a perfect command of both languages as well as a profound understanding of Charbonneau's thought. This difficult task has been entrusted to Christian Roy, a bilingual Canadian scholar specialized in the history of personalism, whose work first established Charbonneau's role in the inception of political ecology. Familiar with his work, he has often visited Charbonneau and has had many discussions with him. No one could be better qualified.

References

Cérézuelle, Daniel. *Écologie et liberté. Bernard Charbonneau, précurseur de l'écologie politique*. ['Ecology and Freedom. Bernard Charbonneau, A Forerunner of Political Ecology'] Lyon: Parangon/Vs, 'L'Après-développement' series, ed. Serge Latouche, 2006.

Charbonneau, Bernard. *La Fin du paysage*. Conception, photographies et légendes de Maurice Bardet ['The End of the Landscape: Design, photographs and captions by Maurice Bardet']. Paris: Anthropos, 1972.

Charbonneau, Bernard. *Tristes campagnes* ['Sad Countrysides']. Paris: Denoël, 1973.

Charbonneau, Bernard. *L'État* ['The State']. Paris: Economica, 'Classiques des Sciences sociales' series, ed. Hervé Coutau-Bégarie, 1987.

Charbonneau, Bernard. *Le Système et le Chaos: Critique du développement exponentiel* ['System and Chaos: A Critique of Exponential Development']. Paris: Anthropos, 1973; Paris: Economica, 'Classiques des Sciences sociales' series, ed. André Béjin and Hervé Coutau-Bégarie, 1990.

Charbonneau, Bernard. *Le Paradoxe de la Culture* ['The Paradox of Culture'], Paris: Denoël, 1965, reissued in *Nuit et jour, Science et*

---

[1] Letter from Lani Niles to the author, 1985.

culture. Paris: Economica, 'Classiques des Sciences sociales' series, ed. Hervé Coutau-Bégarie, 1991.

Charbonneau, Bernard. *Je fus: Essai sur la liberté* ['I was: An Essay on Freedom']. Bordeaux: Editions Opales, 2000.

Charbonneau, Bernard. *Le Jardin de Babylone* ['The Garden of Babylon']. Paris: Gallimard, 1969; Paris: Éditions de l'Encyclopédie des nuisances, 2002.

Charbonneau, Bernard. *Le Feu vert. Autocritique du movement écologique*. Paris: Karthala, 1980, reissued with a foreword by Daniel Cérézuelle, Lyon: Parangon/Vs, 'L'Après-développement' series, ed. Serge Latouche, 2009.

Charbonneau, Bernard. *Finis Terrae* ['Land's End' (in Latin)]. La Bache: À plus d'un titre, 'La ligne d'horizon' series, 2010.

Charbonneau, Bernard. *Le Changement* ['Change']. Vierzon: Le Pas de côté, 2013.

Charbonneau, Bernard. 'Le sentiment de la nature, force révolutionnaire' ['Feeling for Nature as a Revolutionary Force'], *Journal intérieur des groupes personnalitstes du Sud-Ouest (Bayonne, Bordeaux, Pau and Toulouse)*, June 1937, included in a collection of early writings by Bernard Charbonneau and Jacques Ellul, edited by Sébastien Morillon and Christian Roy, *Nous sommes des révolutionnaires malgré nous. Textes pionniers de l'écologie politique.* ['We Are Revolutionaries in Spite of Ourselves. Groundbreaking Texts of Political Ecology'] Paris: Éditions du Seuil, 'Anthropocène' series, ed. Christophe Bonneuil, 2014.

Prades, Jacques, ed. *Bernard Charbonneau: Une vie entière à dénoncer la grande imposture* ['Bernard Charbonneau: A Lifetime Speaking Out Against the Great Deception']. Ramonville Saint-Agne: Erès, 1997.

PREFACE

The heart of the subject

This is where we all are, and this book's author more specifically. Like our industrial world (call it scientific or technological, it matters little since it is all this at once), the so-called ecological reaction that opposes it is everybody's business. Someone who works at Marcoule[1] and reads *L'Expansion*[2] is still liable to lose his job, his cutting-edge industry having suddenly become rear-guard. You can be the CEO of Poclain[3] and still be under threat of expropriation in the public's interest by the bulldozers of the highway, the nuclear plant, the high-speed train, the wide-gauge canal (nothing being on any smaller scale nowadays), a military base or a marina and so on. Let us leave it at that, since this new flood drowns everything along with space itself. Not to mention its varied natural or human repercussions, such as the great leaps backward provoked by a great leap forward, as in Iran. Even if we don't want to know it, we all have over our head the sword of Damocles of the progress – as unstoppable as Progress itself – of the ultimate weapon's spread. We are all dwellers of the earthly *oikos*,[4] whether we consider ourselves ecologists or not.

---

[1]Translator's note: A gigantic nuclear site in the Côtes-du-Rhône wine region of France, operational since 1956, and where the first industrial and military plutonium experiments took place. On 12 September 2011, one person died and four people were injured in the explosion of an oven melting mildly radioactive metal, causing a safety cordon to be set up around the plant.
[2]Translator's note: A French business magazine founded in 1967.
[3]Translator's note: A French family business founded in 1927 by Georges Bataille (no relation to the philosopher) that was once a leader in the field of hydraulic motors for excavators, until it was almost wholly taken over by the US company Case in the 1970s and 1980s.
[4]*Oikos* is a Greek word meaning 'dwelling'.

But more than others, the author of this book has found himself in this maelstrom's heart. Before going into a detailed analysis of the 'ecological movement', he must say why and how he came to be involved in it. When speaking of a human phenomenon, observation depends on the observer, as the object depends on the subject, to the point that he too is entitled to be considered its subject. You cannot know one without the other, as genuine objectivity gives subjectivity its due. And it is reason itself that makes this blood offering to passion.

Thus, at the burning focal point of an age in motion, in progress and in disarray, there is the man who at this very moment thinks it and gives it words. His viewpoint is that of the generation of the crossover, which, reaching the end of its life, is at risk of losing its footing in midstream. As a child of a city smaller than another because it belonged neither to the north nor to the south of the country, open to all the winds of the ocean and of history, he played in its streets without being chased away from them by the first cars.[5] Caught within its walls, he sought to escape it in the vast woodlands that began on the doorstep of the city's last small shops. Since the war reinforced this confinement, he escaped to the countryside where he spent most of his life. If he did not till the earth of Béarn or the Basque Country, he lived on it. Weighing his words, he can say that he has known rivers of water, not of wastewater, and their billowing of algae, not of plastic strips. He lived in the countryside of Montaigne and Balzac, where the roar of engines did not yet cover the growl of mountain streams nor the calls of bells and people. And he saw it change, or in other words, waste away. Can we help talking of war when we now pass in front of those crumbled ruins in the shrubs, having seen them all white, adorned with roses and purple shutters?[6]

While his friends were making their way up to Paris from the provinces – where the occupier had forced them to withdraw, he

---

[5]Bernard Charbonneau, *Le Jardin de Babylone*, Gallimard, 1969 [Paris: Éditions de l'Encyclopédie des nuisances, 2002]. The reader should excuse the author for citing his own works. When one is convinced of having something to say and that one cannot rely on the loudspeakers that now pass for *vox populi*, there is no better advocate than oneself.

[6]Bernard Charbonneau, *Tristes Campagnes*, Denoël, 1973 [Vierzon: Le Pas de côté, 2013].

burrowed into deepest France, which is not French, but Flemish or Béarnese, to build his house there. If he was no peasant, he was rural by choice, and thus as alien to the city as to the country for which any consciousness is a call to the Centre. He took a distance, which gave him some perspective. And silence, times woven from the moments of happiness lavished on the waters by the sun, the shade of oak trees – all allowed him to cross with open eyes this uncertain passage leading somewhere we would rather not know. From his urban adolescence, he knew that what mattered was not the historical sermons blaring out of loudspeakers and cannons, but everyday change as perceived on his street and in his village. And he immediately felt the need to say – and soon to shout – that whatever else was happening was but a dream. But too soon – other people's deafness struck him dumb. In the age of Stalin and Keynes and the Tour Montparnasse,[7] it was out of the question to be saying that there is such a thing as the Great Moulting[8] and that it raises some issues. Still less to write it, save for oneself and a few friends who had more pressing concerns; not to mention doing anything about it, like having some meetings and camps that would scatter under pressure from political and military developments. In the wake of a total war, it is no more conceivable to get anyone to publish a critique of the State than it is permissible to cry: 'Pan is dying!'[9] And

---

[7] Translator's note: Built between 1969 and 1973 in Paris, the Tour Montparnasse was France's tallest skyscraper until 2011.

[8] Translator's note: With the term *Grande Mue*, which he always uses to refer to the main object of his reflection, Charbonneau has in mind something very much like the Great Acceleration described by one of the originators of the notion of an Anthropocene era, characterized (aside from the impact of fossil fuels) by 'a major transition' whereby 'the characteristic habitat of the human species, which for several millennia had been the village, now was becoming the city'. (Will Steffen, Paul J. Crutzen and John R. McNeill, 'The Anthropocene: Are Humans Now Overwhelming the Great Forces of Nature?' *Ambio*, 36, no. 8, December 2007, http://www.ambio.kva.se). Charbonneau came to the same realization from his own everyday experience at an early age in Bordeaux, where he was born in 1910 – the date of the beginning of the incubation period of Stage 2 of the Great Acceleration that officially started in 1945. Charbonneau described it in those terms that very year in a public lecture entitled 'The Year 2000', putting Hiroshima in that wider context. The Great Moulting thus referred to mankind's shedding of Nature as its environment, as it became encased in a Second Nature of its own making.

[9] *L'État*, self-published [Paris: Economica, 1987]. The bulk of *Pan se meurt* ['Pan Is Dying'], partly written before the war, was integrated – once updated – in *Le Jardin de Babylone*.

then, again too soon even though it was already too late, it became possible to publish a few books; but before the crisis of May 1968, it is no use talking of the three hundred thousand French casualties of the automobile, and it is still out of the question to get a critique of growth printed.[10]

This Great Moulting that puts into question the fate of the earth and of man, the author of these lines has gone through it alone, almost without ever being able to share the harrowing awareness of it. No party ever noticed it, be it only a few naturist circles where he felt as much a stranger as in the cathedral of Marxism or Progress. Alone, for having wanted to cry out what was so obvious and to join with other people in the only endeavour that has any meaning on the eve of the year two thousand: *saving nature and freedom*. It takes the obstinacy of a lifetime spent doggedly following one's own path to dare to soberly pick up those two words from the dustbin of history.

***

An ecologist? No, the author has no claim to this title, he does not have his licence. He did not come to ecology belatedly like some intellectuals of his generation: be it scientific or ideological, ecology came to him. And soon, finding himself alone in the crowd, he had to keep a distance. On the subject that obsessed him, an invisible red light had long blocked his path. Then, one fine day in 1970, the green light was given, and he suddenly found himself caught in the stream of the ecological movement; at once within and without, which has always been the condition of man on earth. And he was surprised to find himself among the troops, under the sway of authorities he used to know to be Stalin-friendly and progressive. Taking part in this movement is a bit of a joke for someone who has always asked himself those questions – and a few more. Someone who remembers all the times he was turned down is liable to give in to bitterness and to the wish to claim the inventor's privilege, which is of little interest, even though an invention is not the same thing as a copy. The author had to force himself to forget; if his tone becomes harsher when saying the names of some who bear responsibility, actively or by omission, he should be excused for so

---

[10] *L'Hommauto*, ['Carman', Paris:] Denoël [1967, 2003]. Said critique was eventually published by Anthropos under the title *Le Système et le Chaos*, 1973 [Paris: Sang de la Terre, 2013].

many years of mute fury before the genocide and geocides through carelessness that were being committed under his eyes.

Despite being an academic, he deliberately chose to remain what he always was: one man among others, speaking their language. Neither an artist nor a scientist nor an ecologist nor a sociologist, nor even an organic farmer or a leftist, even in this small green township he feels he is from somewhere else, although a member of the species. He is only an individual, alone as any other when he casts a glance at himself and the universe (I should be saying the environment) that surrounds him. But, strange to say, it is this loneliness he felt for so many years, whose experience is now verified by social authority, which now gives him the right to speak.

The author's viewpoint is thus essentially vernacular and personal. It is not that of a specialist, not even of a specialist of synthesis or subjectivity, of a *bona fide* philosopher or poet, but that of a man who needs air to breathe, water to drink and dive in, time and space to play, silence to sleep or reflect. It is the viewpoint, still rather beautiful, of a country-dweller, who has felt in his bones and his eye each of the blows the land has endured. He speaks of the battle as a combatant, not as a strategist. May science forgive suffering for not being objective enough, for it has a science of its own that is incomparably sharper when it clenches the teeth to preserve sanity and lower the volume. There can be no question of doing a sociology of the ecological movement for someone who is caught up in the situation and the struggle; it is going to be examined in the name of its cause and its principle. What will follow is an auto- (not a hetero-) critique. If it seems hard on 'ecologists', they must understand that it is to help them win their true victory.

* * *

Such a viewpoint has its own method and its own language. What happens around us (again the environment!) is directly perceptible. It is not a detail but a global phenomenon, an ocean whose waves come to die at our feet – when they do not carry us away, which ends the problem. The Great Moulting and its multiple effects can be grasped by a person's sensibility and reason on any given day. Whereas the various specialized methods – Marxist, structuralist or even systems-analytical – enjoying the favour of the average ignoramus will at the most only isolate an impalpable sliver of it. And since, like the situation, the ecological project is a global one, we should not be surprised if our analysis must reject dichotomies, for instance that of sensation and reason, and if we refer to common sense as much as to revelation or lived experience. However, since common sense is suspect of being bourgeois among the intellectual bourgeoisie, it is better to speak of Englishmen's *common sense*.[11] Let us show our cards, something that today is more inappropriate than getting naked. I refer to common sense because I believe that the freedom to think, to reason and to speak belongs to everyone or to no one.

Some will find it strange that an academic would stick to what is known by a normally cultured individual with his eyes open: in this area, his ambition would be to learn nothing that is not already known. Why accumulate still more information? Our society is taking care of it. It is better to unblock knowledge that is stored without our being aware of it in the personal or collective unconscious, and which dark forces are preventing from coming to light. The work undertaken by the author starts where that of specialists stops, with shedding light on the obvious or on 'facts' that no one is considering, because of their being facts. To each his own task: for some to gather, for others to interpret what has been gained and to work out its hidden significance.

So, we are not talking about a sociology or history dissertation on the ecological movement: such a work will always get done because it is honourable and paid. The reader should not expect any statistics, still less mathematical curves; the ones that are useful because they do not seek to prove what is proven in advance are to be found elsewhere. However, if not references, he will find here referents related to his life; we have seen which labels are applied to them. Later, we will have the opportunity to dot both these i's: the alpha and the omega of the ecological movement.

***

To pursue any ends, we must do it in accordance with their means, that is, at the stage of expression, according to their language. The struggle for nature, a fruit of everyone's and everybody's freedom, can only be led in equality (one can see that the author is a left-wing intellectual). It happens that the first one is that of knowledge and speech. Since it is every human's business, not only

---

[11] Translator's note: In English in the original.

that of a distinguished sociologist or biochemist, the demonstration must be made in a language accessible to all. Hence, for other reasons as well, the refusal of the various scientific or ideological Church Latins that ensure the esoteric prestige of a supposedly democratic intelligentsia vis-à-vis the televisual mass that is losing the habit of reading and writing. Since it is our topic, let us remind 'ecologists' about how much of an ecological medium everyday French is, compared to audiovisual media that gobble up folk cultures and kilowatts. But for an Englishman, it will be English, human unity being made of differences. The simplicity of this supple and precise instrument, able to express the finest nuance, puts it within reach of all those who make the effort of laying hold of it. What a wealth of physical and spiritual information accumulated by generations of individuals in so many idioms! A vernacular speech, as deceptive about its depths as clear water, and not – as in some fashionable discourses – covering up shallowness with murk. A popular and learned language, a measure secretly alive with the blood and meaning of an existence whose roots and end elude it in the final analysis. What a contrast with the specialized jargons whose rigour conveys ever-slimmer content. Especially when it comes to human facts. How can we speak of man and his land without breaking the bread of his speech? So, there can be no more question of using learned words than for ecologists of using ersatz materials, the only ones fit for consumption being those polished by use.

Human ecology is the discovery by every man of the upheaval of his daily existence, around him and within him. The author will therefore speak of it in everyday slang, which is not the slang of today, out of fashion tomorrow. In the patois of Flaubert and Jean-Jacques, enriched until now without breaking its tradition, which remains that of the normally educated Frenchman. A dialect as threatened by the howling silence of our deluge as Basque or Breton, and yet even less defended, in spite of the treasures it holds. It is thus in the full awareness of the scandal of its use by an intellectual that the author employs classical and vernacular French to express social reality. But it is the language of the subject in both senses of the term. More inclined to delve deeper into it than to learn another one, he has chosen to become what he is and to show the way to his neighbour. Should the latter no longer understand this lost language, he can always ask a licensed interpreter who will translate it in the Esperanto that is fed into computers. Should this text somehow go through the Big Machine, let us hope it will fry its circuitry.

**

At the flaming red heart of our industrial and military forge, in this twilight of a millennium, *a green light now burns bright: ecological revolt*. Depending on what we do, it will either be a mere social sign leaving the flood of things to follow its course, or *a lifesaving beacon*: the tender, shining sprout of another spring on earth.

PART ONE

Seeds: The origins of the ecological revolt

For a long time, everything happens in the depths, in the silent blackness of the earth's or the Mother's womb, for the drills of human industry are strong and precise enough to penetrate hell. And then, one fine day, the phenomenon bursts forth. May it take the time to grow and mature under the earth's great sun!

CHAPTER ONE

Origins

To understand the movement that has recently come to be known as ecological, we must go back to its origin. But this brief inventory of its sources and its evolution will be limited to the facts that are indispensable to its explanation, which will be emphasized here; the biggest ones not necessarily being the best remembered. One of the received ideas in French ecological circles is that it somehow came out ex nihilo one fine day in May, whereas its gestation goes back to the beginnings of our urban and industrial society, and even of society itself insofar as it has raised in cities and palaces a rampart of stones and sacredness to protect itself more or less efficiently against nature, which was besieging it in full force. Think of the tears shed by Ronsard already on the oak trees of Bercé, which allow the zealots of mechanical progress to invoke the axe as an alibi for the ravages of the buzz saw.[1]

Prehistory of the ecological movement

The love of nature, sometimes going together with hatred for it, is without a doubt as old as *homo* somewhat *sapiens*. But until the

---

[1] Translator's note: The famous Renaissance poet Pierre de Ronsard (1524–1585) grew up near the ancient forest estate of Bercé (5400 ha) in France's present Loir-et-Cher department. It was known for centuries for the quality of its oak trees, but the trades and crafts traditionally associated with them disappeared in the middle of the last century.

modern era, which in many countries only began after the Second World War, it is more a kind of *amor fati* that makes one worship what cannot be escaped: nature is Zeus brandishing his thunderbolts. If the Christian religion has broken this kind of idol – a development with huge implications – the holy or cursed bond between the human mind and the earth was not severed; the Old and New Testaments preserve many traces of the old and new love of nature.

In more civilized societies, be it those of the ancient world, of the Renaissance or of the classical era, the joy of inhabiting the earth manifests in cities or courts among individuals who are sensitive and have a gift for saying it. But nature then is too powerful for their love to apply to what is inhumanly grand about it. As in La Fontaine, when it is praised in poetry, it is especially in its tempered forms, once civilized by country people, and not in the monstrous form of dizzying peaks that call to mind powerlessness and death.

Then, at the dawn of modern times, in the more civilized countries, before they were called developed countries, 'feeling for nature'[2] appears among some particularly sensitive human recorders who have had a premonition of the coming earthquake. This 'feeling' no doubt existed, but was not felt very much and still less was it expressed. It is not in Manchester, the cradle of mechanization, but in the Protestant Geneva of bankers, lost in its countryside, that Rousseau was born. And he has already said everything: all at once Nature and Freedom – unfortunately by adding capital letters. He took the side of the individual and the noble savage against the civilization that corrupts him, and theoretically attempted to build another one on this hypothetical material. But he did not formulate, let alone resolve, their contradiction. And unfortunately, the ecological utopia, in spite of the failures of many revolutions, is still at that stage.

---

[2]Translator's note: 'Le Sentiment de la Nature, force révolutionnaire' was the title of an essay in which Charbonneau first laid out his key ideas about nature and freedom in the internal newsletter of a southwestern branch of the French personalist movement in 1937. This was the kernel out of which his later works on the topic were developed over the years, until the publication of his book *Le Jardin de Babylone* in 1969. This manifesto was first published in book form, along with other early essays by Bernard Charbonneau and Jacques Ellul, in *Nous sommes des révolutionnaires malgré nous. Textes pionniers de l'écologie politique*. Sébastien Morilllon and Christian Roy (Eds.) (Paris: Éditions du Seuil, 2014) 'Anthropocène' series, ed. Christophe Bonneuil.

A whole vanguard of intellectuals and idle rich of Anglo-Saxon and northern countries would stampede onto the path opened by Rousseau and his epigones, Catholic Latin and rural countries trailing at the back. And the pressure of industrial and urban progress would not stop growing. In the United States, where the economy explodes in a pristine space and nature, Thoreau, the naturist, individualist and anarchist Puritan, retires into the woods, refusing taxes and war. Melville worships and fights the great whale – and not only pen in hand. The Englishman Stevenson and the Frenchman Gauguin, far from the cities, their factories and their laws, retire in a Polynesian Eden that only has fifty years left at the most. And following them, a flood of tourist voyeurs – rather than voyagers – rush in.[3]

But society has also neutralized a reaction that was going against its deepest tendency: power over nature and people. It did this by forcing back into luxury (Adam Smith would not have said the 'quality of life') and leisure reserved for the rich what was the expression of a vital need for air and freedom. Nature is thus assimilated to its opposite: to refinements of literature and Art, to the entertainments of princes, to the vacations of lords and then of the bourgeois. The national park for the pleasure or the instruction of the people starts with the garden of the Petit Trianon. The false relationship to the earth of a queen who plays the shepherdess would spread to an entire class. And it persists even in the attempts of a city-bred youth to return to the earth.

In bucolic literature, the struggle for nature finds an outlet even as it is contained. And it is not over – witness the recent success of this kind of books, for it is no longer a few bourgeois, but most of the public that buys them. Romantics counter scientific and industrial rationalization with the spontaneity of nature and freedom, but they need no less than typhoons or the Alps to answer giant cities in kind. And their very outrageousness reassures the class in power, which only sees in it an artist's fancy. The romantic and anarchistic tradition continues in the novels of Giono, but the test of war will do little to confirm the seriousness of the social, pacifist aspect of his work, and the 'no' said to the vanquished army will only end up in a 'yes' to the victor.[4]

---

[3] See Bernard Charbonneau, *Le Jardin de Babylone*.
[4] Translator's note: Jean Giono (1895–1970) was a French novelist who made his name depicting Provence's rural life as a critical alternative to industrial society. His

Pourrat[5] and Ramuz[6] are more modest but they hardly venture out of the literary confines, rustic to be sure, but well-circumscribed, in which fashion has placed them. If the consequences of progress are criticized, as in Ruskin, it is mostly from the vantage point of aesthetics: progress makes ugly; more than life, it is the latter's quality that it jeopardizes. And this aestheticism idealizes nature and the past where, closer to nature, people led a harmonious life.

In the wake of the First World War, Duhamel writes *Scènes de la vie future*,[7] which is a good title, but it rails against the very American horrors of an agribusiness that Giscard[8] now dreams of bestowing to France. Duhamel is just an old reactionary, but he is also an anti-American intellectual, as many ecologists still are.

The thunderbolt of the First World War, which began to dispel the Belle Époque's dreams of progress, put an end to optimistic science fiction as practised by Jules Verne. The ancestor of such pessimistic futurism was, strangely enough, the cautious Anatole France, who, in his *Penguin Island*,[9] prophesies the end of civilization in generalized terrorism and explosion. This future, already ambiguous in the work of H. G. Wells, whose optimism was turned to pessimism by the Second World War and Hiroshima, becomes altogether black in René Barjavel's *Ravage*.[10] As in the American science fiction that greets the advent of the atomic age, the world of the future is identified with the end of man and his freedom. And

---

frontline experience in the First World War made him a standard-bearer of pacifism, who came to be criticized for a too-ready acceptance of German Occupation during the Second World War.

[5]Translator's note: Henri Pourrat (1887–1959) was a French regionalist writer and folklorist of Auvergne.

[6]Translator's note: Charles-Ferdinand Ramuz (1878–1947) was a Swiss writer known for his depictions of rural life in the canton of Vaud.

[7]Translator's note: French writer Georges Duhamel (1884–1966) is largely known for the international bestseller *America: The Menace; Scenes from the Life of the Future*, translated by Charles Miner Thompson (Boston, New York: Houghton Mifflin, 1931), a travelogue of contemporary American society as a troubling harbinger of a coming technological mass society of endless production and mindless consumption.

[8]Translator's note: Valéry Giscard d'Estaing (born in 1926) was president of France (centre-right) from 1974 to 1981.

[9]Translator's note: *L'Île des Pingouins* (1908), a satirical fictional history by this Nobel Prize–winning French author, translated by A. W. Evans as *Penguin Island* (New York: Modern Library, 1933).

[10]Translator's note: A 1943 post-apocalyptic classic of science-fiction literature, set in France in 2052 after electricity suddenly disappears worldwide, translated by Damon Knight as *Ashes, Ashes* (New York: Doubleday, 1967).

the only chance he has left is starting anew on a devastated earth or on another planet.

In this review of the now-forgotten ancestors of ecology, we must give a special place to Aldous Huxley, about whom one wonders why he is not, instead of Illich, considered the father of the ecological movement – Jean-Jacques being its grandfather. With him, the novelist steps back behind the thinker dedicated to defending the freedom of the individual. Far more than the work of so many 'committed' *littérateurs*, this one is committed to facing the realities of its own time. In 1980, *Brave New World* has not aged[11] – something that can hardly be said of so many Marxist or Stalinist writings from the last war's aftermath. But the sceptical intellectual's acute critique ends up in a mere return to Eastern spiritualities of detachment and other sacred drugs. Once again, the therapy pales beside the seriousness of the diagnostic.

Nevertheless, without openly addressing the state of society yet, revolt is simmering against a progress paid with human upheavals and many setbacks. It is rooted in the unconscious of the people that is then paying the cost for the use and purchase of the first machines, but also of peasants and craftsmen condemned from the start and then of intellectuals whose values are just as threatened despite their belief in social progress. And we find nature again as much on the Right (where, since Burke, its demands have been contrasted with the Left's revolutionary claims), as on the Left, where capitalist development is sometimes denounced in the name of an anarchist ideal, or of a Spartan one *à la* Babeuf,[12] of freedom and equality in austerity.

---

[11]Translator's note: This novel's French translation, the same year the original came out, had confirmed Bernard Charbonneau's original intuition of the role of technique, which he later asked Jacques Ellul to develop into a book that his friend finished in 1950. Ellul's breakthrough as a prophet outside his own country came with the publication by Knopf in 1964 of its English translation as *The Technological Society*, due to the discovery of the original French book, published in 1954 as *La Technique ou l'enjeu du siècle*, by Aldous Huxley, who promoted it in the United States as it proved what he had tried to say in *Brave New World* in 1932 – coming full circle, unbeknown to him. Ivan Illich, on the other hand, has stated publicly that he considered Ellul his master as a social critic from the time he came across his book's translation. Thus, Illich's status as a father of the ecological movement is literally a derivative of Huxley, by way of Charbonneau and Ellul as the actual line of transmission of the novel's insights.

[12]Translator's note: Gracchus Babeuf (1760–1797) is often seen as the first revolutionary communist or anarchist before these words existed.

Like today's ecological movement, the return to nature manifests itself in the new generations of urban youth, for whom it is a physical and spiritual need. An English general[13] exploits, in the interest of morals and the army, city children's wish to play Indians. The Scout movement remains infantile and only ends up providing a cadre for the final offensive of colonial imperialism.

As for the *Jugendbewegung*,[14] it will extend to the end of adolescence, up to the gates of barracks and trenches. It is born of the mists – worse than the Brocken's – of the witches' cauldron[15] of the chemical industry and German growth, right before the storm of the First World War. But under the impact of events, the *Jugendbewegung* will split between a right wing and a left wing, the latter promptly annihilated by the zealots of Blood and Soil. And after the armistice, all that will be left of it in the north will be peace-loving naturist and vegetarian, hygienist and nudist groups, which practice sun worship. Playing the role of a fixation abscess for the grievances of a marginal opposition, this naturism will be no threat to the reboot of the second industrial society, which will encyst it in the guise of harmless, colourful sects.

As a deep but merely instinctive reaction, the upsurge of a youth in rebellion against all the chains that would tighten around it, the naturist revolt has all the strengths and weaknesses of childhood and adolescence, incapable as it is of going beyond them to access the consciousness and responsibilities of adulthood. We will find again in the ecological movement the vices of the virtues that have prevented the naturist revolt from openly expressing itself and acting on a wider stage.

A great silence

Twice in a row, total war put off the emergence of a naturist revolution (it would only be called ecological later) by giving a new boost to industrial development. And after the second global

---

[13]Translator's note: Robert Baden-Powell.
[14]Translator's note: 'Youth movement' in German.
[15]Translator's note: The Brocken is the fog-shrouded highest peak of northern Germany in the Harz range, and the focus of the witch legends taken up in Goethe's *Faust*.

conflict, this boost is on the same scale as its mobilization and the destruction it brought about. The world's peoples dedicate themselves entirely to reconstruction and the war's shortages confer a magical aura to comfort and cars; as for *Blut und Boden*, they have had their fill of it. And the benefits of progress are still prized wherever underdevelopment and a certain kind of socialism prolong destitution and shortage. The red light of Hiroshima that bars the entrance to industrial paradise seems to have gone unnoticed, even though it imprinted its invisible scar on a generation. But this was only one more reason not to know that, with the Great Moulting, life and death were at stake.

More than in the English-speaking world, in Europe and especially in France, a great silence reigns, covered by the noise of bulldozers and the intelligentsia's still Stalin-friendly discourses. The war that unleashes technological progress hardly lends itself to the critique of a society that mobilizes its intellectuals for a more urgent task; while machines improve, minds regress. And after 1945, more than in 1918, the war's after-effects take a long time to fade. Schizophrenia is at its apex. Just at the moment when human power overwhelms the earth, reflection about society and political action lose their way in noise and fury. Capitalism and communism commune and compete in the same worship; to catch up with the United States and overtake them in the rush towards absolute production–consumption, such is the watchword of their worst enemies. While in the West politicians twist and turn and businessmen reap the profits to invest them with the State's help in the equipment to make even more, in the East it is another Peter the Great who handles this task by wielding the knout. The first H-bomb may explode in Bikini; this publicity spot is barely worth naming panties after it. The Soviet bomb could follow and megatons multiply; that device was even less real. England and France only had to follow suit – and then China and all the rest. Besides, the hopes raised by peaceful nuclear energy made people forget its dark side. More pressing events caught all the attention, such as the military's wanderings – unplanned by contrast – in Indochina and Algeria, punctuated by the Gaullian rooster's clarion calls for imperial unity at first, and then for decolonization. Ideology and politics dream on amid the noise and fury of a bloody spectacle identified with Action and History, but its function really seems to have been to distract from life's all-too-abrupt transformation and deeply repressed anxiety at the prospect of atomic death.

The politicization that followed the war resulted in emptying politics of its contents, at least if the latter consists in realizing on earth man's desires and dreams. Especially in Latin countries, any action not strictly confined to the private sphere is reduced to the seizure or defence of political power by parties and politicians specialized in manipulating the masses. No matter the ends or even the program that is often only smoke and mirrors, if we win, we will see then what we are going to do.

While material upheaval accelerates, political thought is stuck in ideologies, liberal or Marxist, dating from the pre-industrial era. As the Right manages growth, the Left, paralyzed by the intransigence of the Communist Party, is condemned to merely verbal opposition. And wherever it takes power, it contributes to development by perfecting a system that is a bastard of planning and market, one of nationalized free enterprise, which makes it possible to avoid crises and provide jobs for a while.

And the intellectuals, specialists at criticizing, are also absent. Those on the Right, who are technocrats rather than novelists, are in favour of economic progress, and it is not from the nuance-filled analyses of Raymond Aron that we may expect a critique of its nuisances. In France, Keynes sums up Science. Profitability, production and productivity, full employment – these watchwords do not express a side of reality, they reveal the whole of it. In the West as in the East, airplanes and dams are not means but obsessive ends; and it is in vain that the resistance of the mountain people of Tignes (Savoy) will be a reminder that to produce kilowatts you must drown villages.[16]

The left-wing intellectuals who rule over Culture if not power, like most Frenchmen, mobilize against a Hitler reduced to ashes. Stalin, then alive and well, holds much less of their attention, and it is only after his death that they will admit to the Gulag. Marx having forever defined any social criticism, they participate in the mass of the second Rome or stand about at her doors. They sing the praises of work and the workers, of beautiful Soviet machines, being more discreet only when it comes to the red bomb. Or they escape into aesthetic and surrealistic games. But as for the coming

---

[16]Translator's note: The building of this dam between 1948 and 1952 was the occasion of early French protests against industrial encroachments and their effect on the environment.

of the year 2000, on this topic the vanguard lags behind as much as the average Frenchman. Burnham's discovery of technocracy is viewed as a CIA diversion. J. Ellul's big book on Technique,[17] sharply criticized by Professor Duverger,[18] will have to cross the ocean to find an audience. Where are the ecological notables of that generation? Illich is only a progressive Christian who is introducing psychoanalysis in convents.[19] Edgar Morin is a Communist Party militant, and Alain Touraine is writing a history of work inspired by Communist orthodoxy. All experts agree. Mr Mansholt then gives his name to a plan to liquidate Europe's countryside and peasants. Professor Dumont does the same on behalf of the Monnet Plan; it is only later that he will denounce the evils of the 'Green Revolution' in the Third World.[20] Where are Messrs Peccei and Massé of the Club of Rome during that period? At Fiat or at the helm of the Ministry of Finance and Planning. When the *Trou des Halles* is inaugurated, Senator Bonnefous is in favour of an expressway on the quays of the Seine.[21] And the author of the 'socialization of nature'[22] identifies its protection with the kind of regional planning for the Aquitaine coast that would have turned it into a tourist suburb[23] and so on.

---

[17] Jacques Ellul, *La Technique, ou l'enjeu du siècle*, A. Colin, 1954 [Paris: Economica, 1990, 2008].
[18] Translator's note: Maurice Duverger (1917–2014) was France's foremost post-war academic political scientist, internationally known for his work on parties.
[19] Translator's note: Charbonneau seems to be confusing Ivan Illich with his friend Gregorio Lemercier, also based in Cuernavaca as prior of its Benedictine convent, which was shut down by the Vatican in 1967 as a result of its embrace of psychoanalysis that was more or less anathema to the Roman Catholic Church at the time. A 1962 article on this affair in the newspaper *Le Monde* may have been the source of this misapprehension, aside from Illich's association with Erich Fromm in the same Mexican city. (I thank Daniel Cérézuelle for this information.)
[20] Translator's note: The agronomist and sociologist René Dumont (1904–2001) was France's first ecologist presidential candidate in 1974.
[21] Translator's note: The site of Paris' historic marketplace (Les Halles) was left as a gaping hole (*trou*) for several years after the dismantling of its buildings in 1971, until this eyesore was turned into the central hub of the city's metro system.
[22] Philippe Saint-Marc, *Socialisation de la nature*. [Paris:] Stock, 1971 and 1975.
[23] Translator's note: From 1967 to 1988, the Mission interministérielle pour l'aménagement de la côte aquitaine (MIACA) was in charge of providing the infrastructure for the development of France's southwestern Atlantic coast. Bernard Charbonneau and Jacques Ellul were at the forefront of grassroots resistance against its plans.

One would have to enumerate all the notabilities, today more or less converted to ecology, who have wielded influence or power at that time and who could have at least raised the issue and given another face to development. Aside from a few isolated individuals reduced to marginal publications, one can only cite D. de Rougemont, belatedly discovered by the ecological movement.[24] If it is true that only fools do not change, this holds only if you don't change as a band at a signal. And when you have exercised responsibilities as grave as Mr Mansholt's, as long as you publicly acknowledge your (sometimes deadly) mistakes – which some have done, but rather discreetly! And yet, I do not question the good faith of those public figures; perhaps I would have done the same in their situation. At a time of uncertain and changing truths, the sincerity of the individual caught by the social torrent is only too great; and the higher one is placed, the more difficult it is to break free of it. To cross the stream of history, one waits for the green light. How do you oppose it when, from the *Figaro*[25] to the *Nouvel Obs*[26] – let us not even bother with *L'Huma*[27] as it dreams on unaffected by the times – speaking out publicly against the threat weighing on mankind and the earth is out of the question? One might as well give up on any career and action and endure that modern version of damnation: being alone.

Thus, the costs of development at any price, dashing ahead like a crazy horse under the whip of capitalist competition and the State, are all the greater for being neither controlled, nor even expressed. The car maims and kills more than the numbers killed in the Algerian War, amid general indifference. Rivers get polluted, landscapes are being erased and houses crumble without the media or the public

---

[24]Translator's note: The Swiss essayist and diarist Denis de Rougemont (1906–1985), mostly known in the English-speaking world for his 1939 classic *Love and the Western World*, translated by Montgomery Belgion (New York: Harcourt, Brace, 1940), was a theorist of French personalism and a lynchpin between its different wings. A pioneer of European federalism, he always remained faithful to the movement's pre-war roots and close to the initiatives of its Bordeaux School, e.g. as co-founder of the ECOROPA group.

[25]Translator's note: France's main right-wing newspaper.

[26]Translator's note: Founded in 1950 as *France Observateur*, *Le Nouvel Observateur* (as this weekly magazine of the caviar Left was renamed in 1964) adopted its old nickname *L'Obs* as its official moniker in 2014.

[27]Translator's note: *L'Humanité* is the French Communist Party's daily newspaper.

registering: Frenchmen will only notice the rubble on their doorstep when it appears on the TV screen on *La France défigurée*.[28] Besides, France is still disfigured. Except for some isolated individuals, be they intellectuals or victims, the hugest phenomenon to have ever arisen in history remains, for all intents and purposes, without effects. It is happening outside of the space-time continuum, as it were. The environment is not an issue for those being environed, the quality of life being exhaustively defined by its quantity translated into statistics as truths are in surveys. The Great Moulting *is developing* on its own, out of the collective mind, as an avalanche hurtles down the side of an uninhabited mountain. More than a social or even a biological fact, it is a geological one.

And yet, even if we may have different opinions on the topic, one thing is certain: its size and, consequently, its importance. But biology teaches us that for one positive random mutation, there is an infinity of negative ones. And if man is under the obligation to master nature, it is all the more necessary that he should recognize the necessity of mastering his own nature.

---

[28]Translator's note: Public affairs television program about environmental issues that aired from 1971 to 1977 and again briefly in 1995.

CHAPTER TWO

Ecology, Year One

The emergence of the 'ecological movement' has been exceptionally rapid. Only twenty years ago, even in the United States it was unthinkable. The exponential development of everything: hydrocarbons, goods and jobs, *hence* of sea baths and river swimming, of landscapes and happiness, of freedom and equality and so on was the revealed truth of all the world's industrial regimes and of those called to join them. And the economy's juggernaut stormed straight ahead, crushing trees, walls, and sometimes people. Because it was so, it had to be so.

In this vacuum, the explosion of the second industrial society was accelerating: that of oil and electronics, and soon of the biochemical and nuclear industries. But as progress develops, we must pay more and more for benefits that get slimmer and slimmer. If at the beginning the plane reduced the crossing of the Atlantic from six days to twelve hours and the switch from the propeller to the reactor reduced it further to seven hours, for a much bigger investment, the Concorde only reduces it (if we include formalities and the trip to the airport) from seven to five hours. At this rate of development, we may imagine a super-Concorde that will manage to save five minutes by mobilizing all the nation's resources. The same thing goes for the progress par excellence – that of medicine. Initially, it saves at relatively little cost millions of young people or adults from smallpox and then tuberculosis, but pushed too far for a price in equipment that is ruinous for hospitals, it is now only good for prolonging the agony of the dying. Thus, little by little, the drawbacks of the benefits of progress become more perceptible. Furthermore, developed countries, especially their bourgeoisie that profited the most from its benefits, now only have its evils left to show for it.

And so, it comes about that one fine day we realize that 'development is not growth' (you may reverse that formula of 1970 'growth is not development', it doesn't matter). In other words, the divine ascent of the curves has a cost, but since it remains sacred, you have to say it without saying it; the day progress condemns you to backtrack, call it 'quality of life', 'an alternative development model', or 'post-industrial society' at a time when everything becomes industry.

Where the thesis, that is, science and America, generates its own antithesis

The first ones to point out the damage in other than literary terms were scientists: naturalists and biologists, ethnologists, more rarely physicians, forced to acknowledge them in their specialties. Fortunately, science has its share of humanity, its knowledge is not just objective, it also has an interest in its object: love. It is naturalists who invented ecology. They were compelled to admit that nature was no longer the same invincible and eternal object Audubon and Haeckel had known. Its study was now inseparable from that of its degradation by man; to know nature was to know its disappearance, the disappearance of the most fragile species, which are often the biggest and most beautiful: such as the whale, today a symbol of nature for ecologists.

Even before the 1960s, Thoreau's descendants denounced the ravages of agribusiness and pesticides, and Rachel Carson wrote her *Silent Spring* that woke up American opinion. In France, the director of the Muséum national d'histoire naturelle, Roger Heim, nowadays somewhat forgotten by the 'ecological movement', tried to formulate the problem in books and articles and, as early as 1957, led the defence of the forest of Fontainebleau threatened by the highway. Jean Rostand joined in his work the defence of freedom and that of nature. And Professor Dorst wrote *Avant que nature ne meure* in 1965.[1] In 1964, Pierre Pellerin led a campaign

---

[1]Translator's note: Jean Dorst, *Avant que nature meure, pour que nature vive* (Lonay: Delachaux et Niestlé, 2012), this new title meaning 'Before Nature dies, so

against pollution and for threatened species. In 1945, naturalists successfully pressed for the creation of a Conseil pour la protection de la nature, the maintenance or the extension of nature reserves like that of the Camargue and the institution of the national park of the Vanoise (1963). But their action, like that of the Ligue pour la protection des oiseaux contre les chasseurs[2], remained an endeavour of specialists dedicated to the defence of nature in the strict sense. And the few enclosures they managed to obtain from development's planners mostly served as an alibi for exploitation – any which way, any time, anywhere – of the rest of the national territory.

Following Haeckel, naturalists were interested in ecology. It had taught them that everything is related in life: the environment and its inhabitants. That living things belong to an ecosystem: a herbarium, a steppe or a blue planet, where each species has its niche and contributes to the equilibrium of the whole, and that should it be broken by any chance by the disappearance or the measureless growth of one of its factors, death restores the balance. However, more prolific and powerful than any other, the human species tends to disturb this concert. But since its action was long slow and short-ranged, its weakness inclined it to wisdom, and balance was eventually restored: the forest was succeeded by the bocage or the treed moor. The harmony of landscapes is nothing more than the sign of this precarious accord of man with his environment, just as sewers and dumps signal its rupture. And the greyish patches of vacant lot and concrete are but the sign of another leprosy, this one interior, which gnaws at societies and individuals.

However, if the growth of his means has allowed man to suppress suffering and death, it has unleashed imbalance on the entirety of continental space-time and even of the ocean, the source of all life; what is at stake is no longer this or that ecosystem, but the one constituted by the earth. Sooner or later, the curve's ascent towards the absolute will hit upon the limits of that sphere. We may imagine other energy sources than oil if we pay the price; it is going to be a whole lot more difficult to produce an ersatz for the square meter.

---

that Nature lives', was one of the first books in French to openly address the threat posed to nature by human activities.
[2]Translator's note: 'League for the Protection of Birds from Hunters'.

Ecology is a godmother to the revolt against industrial society. But this term, 'ecology' or 'ecological', that is now part of the public domain, is not without ambiguity. From now on, thanks to the media, everything is becoming ecological (or turning green): a demonstration on bikes on the Place de la Concorde, a day of sunbathing inaugurated by a minister and the Antenne 2 network and any bearded assistant at the Muséum national d'histoire naturelle. A label was needed to refer to such a new phenomenon, and there can be none without Greek or Latin; otherwise, it is not a proper label. By becoming vulgarized, ecology has lost its scientific meaning to take on another one, a magical sense that draws its prestige from an esoteric knowledge reserved for a few magicians.

It is not for nothing that this label has been applied to the reaction against progress at any cost. It was too complex and disquieting. It had to be simplified, explained, reintegrated into the society it threatens by relating it to the latter's highest authority: science. Ecology being in the strict sense of the term the scientific study of natural ecosystems, a specifically human reaction was thus reduced to nature and the methods for approaching it.

Now that the damage is done, it seems difficult to refuse a denomination that has passed into the public domain. And yet, it would have been better to refuse a term now applied to human societies as well as plant and animal ones. As we will see later, keeping to ecology means forgetting that man has a mind, just as ignoring it means forgetting that he has a body. The ecological level is but one plan of attack – there are other planes, the psychological, the social, and so on. But if ecology is not the alpha and the omega, it is a point of departure, unlike those political mythologies that are as alien to the nature of the earth as to that of man and the problems of his time.

Let us say a word or two about 'the environment' brought from America to Europe by some other Columbus. Miracle! Durand,[3] who until then was suspended in mid-air, wakes up somewhere on earth. Looking around him, he discovers he has an environment; but if he acknowledges being environed, it is in the way of a besieged person whose space is closing in and tightening. This situation in space may be a good thing, provided the word is not substituted

---

[3]Translator's note: The French everyman, like the English 'Smith'.

for the thing. This term 'environment' makes it possible to give a doctoral air to a common-sense recognition that living things do not exist in a vacuum, but at the centre of a space – amid a milieu. It should not be made into an abstraction and a special category with its own ministry. If it was not an alibi, it should be dealing with everything: we might as well call it the Ministry of Space-time and allocate to it the totality of the budget, which is far from being the case.

To the naturalists must be added the ethnologists. They too were interested in a threatened species: the societies long said to be primitive in relation to Western progress. Ethnology allowed intellectuals who were ill at ease in their developed civilization to take refuge in others still in harmony with nature and the gods, under the pretext of studying them. In the name of science, they could seriously play Indians and be paid for their work. But from Lévy-Bruhl to Lévi-Strauss and his heirs, they lost the good conscience that made them pin on traders and missionaries the ruin of those cultures they loved. And Théodore Monod, and then R. Jaulin went so far as to put into question that other imperialism that had invented and financed their discipline. If their predecessors had been able to entertain illusions about the perennial status of these social microcosms that enabled them to go back to human origins, they saw them decomposing when they were not already annihilated. They sought in them different cultures resistant to time, and they had to record their death throes. How could they study them without wondering about their ruin and its cause? Unless he remained a passive observer who was content to put them down on index cards, that is in their tombs, the ethnologist felt caught in an unbearable dilemma: either helping those survivors to integrate, in other words to disappear, or managing reserves forbidden to the public, as are all true nature reserves. In both cases, to preserve life, it was put in a bureaucratic freezer. And the sincere ethnologist only had one way out left to him: going to war against the cultural levelling of the earth. But on that day, it is no longer the faraway Bororo that must be defended, but the Indian of Berry or even of Paris.

Interest in nature or the 'noble savage' is a function of the 'development' of society and the individual. It is greater in urbanized nations of older vintage and, in those countries, among their rich and cultured bourgeoisie. On the other hand, concern for nature will be less keen among peasants and peoples who have just torn themselves from the land. Hence the misunderstanding that separates ecologists from the Third World whose sympathy they seek. To get out of destitution, the latter puts its trust in industry, as in those two other products of the West: Marxism and nationalism, just as the West begins to have doubts about them. The peoples of Africa and Asia claim their right to economic – and sometimes even demographic – growth at the Stockholm (1972) and Bucharest (1974) conferences. But the fashion of ecology will reach their bourgeoisie sooner or later, at least wherever their regime does not copy that of the USSR.[4] In this case, no problem, since Marx solved them all at a time when the earth's resources appeared inexhaustible. In the USSR, that other new country, the development of nature was the engine of the regime and of the plan. Since Russia has at its disposal a sixth of all continental space, it still has opportunities for exploitation that are far larger than those of most countries, particularly those of Eurasia's little cape, and only a few samizdat addicts are interested in pollution in Lake Baikal and the Volga. Thus, ecologists must expect to meet with just as much resistance on their left as on their right.

Public opposition to industrial society was thus born in the most advanced one: in the United States, and within the United States, in the richer states like California. In the pioneer era, progress had not taken place there as in Europe, in a space occupied by cities and countryside; it had been able to directly exploit a nature whose capital was intact. No human obstacle, but for a very loose seedbed of Indian tribes, had checked the advance to the Pacific. No local tradition had stood in the way of the rationalization of the economy. But the spirit of freedom inherited from Protestantism remained strong, nurturing, along with a deep layer of bad conscience, interest in this nature and the Indians that America was destroying. Hence, as early as the second half of the nineteenth century, taking advantage of available spaces, the creation of national parks and Indian reserves that for a while gave the illusion that the contradiction between progress, nature and Indians could be resolved. But today we know that the function of these institutions, crushed by the stream of tourists, has less to

---

[4]This was written before the Iranian Revolution, which may be thought of as a reaction to the traumas inflicted by exceptionally brutal development.

do with preserving small islands of nature or of different societies than with recycling this refuse for the benefit of the economy and of society as a whole.

The Puritan tradition that is at the origin of the development of the United States has also maintained sects that refused to go along with progress in the name of a Christian ideal of austerity and morality. The best known is that of the Pennsylvania Dutch, who shun the car and the tractor and keep to their traditional attire, as Indians of another, more prosperous kind, whom tourists go to visit. There also remains some trace of this ideal among the Quakers, who favour a simple, non-violent life, despite their financial efficiency: the term 'Friends of the Earth' is reminiscent of the one that Shakers use for each other. Furthermore, the empiricism of Anglo-Saxons is interested in the immediate environment, and their social sense drives them to band together to defend it. The contradiction between a spirit of freedom that is more demanding in practice and daily activity that is ever more massive and organized was to lead the American people to react before others. The persistence of the American tradition got combined with its being put into question by the closure of space and various wars to generate a sharp countercurrent at the beginning of the 1960s; but it too belongs to the stream.

I do not have to repeat here a description of the hippy phenomenon that has been done twenty times over. I will only recall to what extent it is the inventor of the slogans and uniform that May 1968 put into fashion in France. To the thesis answers the antithesis: long hair to short hair, grime to hygiene, non-violence to force, discomfort to comfort, poverty to money, sex and drugs to morals, nature to the city and so on. Aside from this, local defence, traditionally active at the level of municipalities and neighbourhoods, takes a new shape. Ralph Nader criticizes General Motors cars, he campaigns against nuclear plants, pollution and products that are hazardous for consumers. The hippy movement may decline; the various lobbies for the defence of nature remain active. They can rely on a legal system that is more liberal and better respected by the State than in France. They would thus slow down the development of the nuclear industry and induce Carter to advocate for energy savings. But, despite the crisis, the difficulties he meets with in implementing his policy show how much this countercurrent remains incapable of inflecting a social process geared to expansion.

The Green Light in France

In France, the ecological movement emerges all at once, at the signal given from above, as though society and its government had wanted to make the first move. We may therefore speak of a green light for this other reason.

A few events announce it in the years that precede the crisis of May 1968. The year 1962 saw the creation of the Association pour la protection contre les rayons ionisants, and the APRI campaigned against X-rays. In 1964, organic farmers founded *Nature et Progrès*, and somebody who took part in its modest beginnings may be allowed to say how much it was ignored at first by the public and the official agrochemical sector. In 1966, taking advantage of the fashion for small farmhouses, Maisons paysannes teaches the public that the defence of the rural habitat is no less worthwhile than that of historic castles. The first defence committees start to become active in the regions most at risk from the industrial tide: in Alsace and the Rhône corridor. But it is only after the events of May, in October 1968, that societies for the protection of nature gathered in the Fédération française des sociétés de protection de la nature.

The silent malaise fostered by unbridled development bursts forth in broad daylight with the revolt of May 1968. It occurs in the part of the population that is most sensitive and alive, where the need for happiness and freedom is felt most keenly, amid the youth of the high schools and faculties of the city par excellence. A surprising, incomprehensible crisis if we refer to the ideologies that prevailed at the time, which may remind us of another student riot, less spectacular but more murderous, at the end of the last century. Like the earlier one, the May riot is born of boredom, of the need to break up an order both too durable and too heavy, to change life and the street, and not just a political constitution that is only good for reinstituting or regenerating society. The May crisis is a phenomenon of rejection, this time of school, work, morals. But this rejection remains unconscious; the happy revellers mix the old Marxist or anarchist formulas with surrealist or situationist novelties. And the crisis becomes fixated on teaching when its cause is the torrential weight of the economy and the State. No more than older people, even the young could not grab such a bull by the horns; this would have entailed turning the festival into work and sacrificing what made it so beautiful: freedom. Hence a channelling into mime and spectacle. This theatrical release of tension was tolerated by the powers that be only so long as it did not threaten the social order; the day it seemed to do so by getting close to the factories, the bourgeoisie and the army put an end to it with the tacit agreement of the Communist Party. There was yet another educational reform, and the machine started to hum again to make up for the time lost.

As a fit of collective hysteria, a neurotic signal spontaneously issuing from the depths – unlike organized political revolutions, the May crisis seems abortive. But for the sexual revolution – though the crisis of the family, as a product of industry more than of freedom, had preceded it – it realizes none of its dreams of abolishing the factory, barracks and school; at the most, it changes fashion. It only innovates on one point, thanks to the situationists: the critique of industrial society that had been blocked until then by the authority of Marxism and liberalism. It is not Edgar Faure's reform, but the ecological movement that will be the lasting fruit of the failed May spring.

This is due in part to a man who is no longer with us today: Pierre Fournier, journalist at *Charlie Hebdo*. As for the young of his generation, ignorant of the past, the 'ecological problem' had been a genuine discovery, which cannot be said of the old belated converts. Shouting in a left-wing paper the ravages of development and its wastage, against the grain of the mythology of material progress, required courage and the freedom to stand in opposition to his own milieu. Fournier said it in the language of the young, but without mincing words. He did not announce a brighter future, but the end of time, perhaps sensing that his own time was counted. Fournier's death is a heavy loss for the ecological movement, for such naturism is hard to see who today can make it keep, along with its intransigence, a seriousness that is not necessarily that of numbers.

Finally, in 1970, decreed 'Year of the Protection of Nature' by the Council of Europe, the green light is given by the highest political authorities. It will be talked about; it is an order. Following Nixon, President Pompidou, whose name is inseparable from the most brutal phase of expansion in France, takes up the watchword that has come from the United States. Chaban-Delmas, the head of government, elaborates on the theme. The civil service follows suit before the nation. And the Aquitaine *igame*[5] complains of not being followed by the notables of the coast that gets developed under the pretext of the protection of nature. Indeed, for the various technocrats and concrete-pourers, the one is really the other (*viz.* the colloquia held at Royaumont on the theme of 'development and nature'). Since in this country everything ends up with a ministry and its offices, one is created for the environment, which has now become 'quality of life'. As for quantity, all the other ministries and 99.9 per cent of the budget were still allotted to it. Until 1972, the effect is recorded but not the cause. The report of the MIT[6] for the Club of Rome will ask the question of growth, quantity as embodied by the computer alone being entitled to denounce the effects of its cult, and Mr Massé (of said club) the shortcomings of its plans.

For the French government, the theme of the conservation of the most concrete realities there can be – nature and the lifeworld – is, as it were, abstract, overlaid on the real issues, which are economic and political. It is but a concern that belongs to a minority of researchers and functioning lunatics; it may speak well of the culture of those who make these kinds of sounds, but remains out of touch with action. Ecology is merely a kind of aesthetic gadget, a green touch that looks good in the verbal landscape. Since growth remains a taboo for the serious people in charge of the balance of commerce and employment, the avalanche follows its course unaffected. But a verbal foam – pollution! – is added from now on to that vomited by pipelines and smokestacks. Life goes on ... Air and water thicken, dumps rise, highways start, oak trees fall. It is the Year of the Protection of Nature, or Arbor Day.

Gushing out of the valve half-opened by the head lock-keeper, a green flood, largely spectacular and verbal, suddenly invades France. While parking lots and factory bread replace the countryside and its pies, desert islands and rustic bread invade the metro's corridors. No self-respecting polluter can fail to also be the sponsor of a nesting box on some sand bank (the reader may fill in his own

---

[5]See the former IGAME prefects. Translator's note: This is an acronym for *inspecteurs généraux de l'administration en mission extraordinaire*, regional prefects responsible for several French departments with both civilian and military authority from 1948 to 1964, when they lost their common name *igame* along with the latter. Charbonneau's use of the term is therefore anachronistic.
[6]Massachusetts Institute of Technology.

examples). Ecological associations multiply. Only in 1970, Jeunes et Nature, the French section of Friends of the Earth, Journalistes et Écrivains pour la protection de l'environnement, the Société française d'écologie and so on. The media rush: information, or rather spectacle, is their business. And nature, this avatar of culture, is a photogenic star. The protection of nature and the lifeworld makes its appearance in the press in an 'Environment' section, where the sensible ecologist promptly gets a job. Publishers launch ecological collections from which they expect a windfall, and rustic novels are bestsellers, provided they do not sadden the reader by dwelling on the disappearance of those landscapes and oh-so-endearing peasants. In 1971, television screens were polluted by the garbage dumps of *La France défigurée*, a TV program which remains discreet when it comes to denouncing the ravages of agricultural policy or official development. Let us leave aside the 'underground' press and the likes of a *Gueule* too widely *ouverte* to really scare serious people.[7] Young people must have an outlet.

Ecology is a discourse, and back then it belonged to the avant-garde. Therefore, if business circles, unions and parties remain guarded, the chattering classes come over to it. But it took them a year or two to understand, and for the various preachers always chasing after their young flock, at least three or four. In 1973, you are no longer in the mainstream of current affairs or of youth if you do not talk about ecology or even ethology. Edgar Morin is coming back from California, where he has discovered a *Paradigm Lost* (wink wink); in other words, that man belongs in part to nature.[8] The director of *Esprit*, whose founder denounced 'the little fear of the twentieth century', is panicking at the prospect of the twenty-first century.[9] From being a scandal, the critique of

---

[7]Translator's note: *La Gueule ouverte* ('The Gaping Maw' – like a dead person's) was the irreverent radical ecological newspaper launched by Pierre Fournier in 1972 as a spin-off of *Charlie Hebdo* (some of whose star graphic artists – like future terrorism victims Cabu and Wolinski – were prominent contributors). It lived on until 1980 after Fournier's death on 15 February 1973, announced in its fifth issue.
[8]Translator's note: Philosopher and sociologist Edgar Morin (born in 1921) published his *Journal de Californie* in 1970 and then *Le paradigme perdu: la nature humaine* in 1973.
[9]Translator's note: Jean-Marie Domenach was by then the director of the review *Esprit*, founded by Emmanuel Mounier (1900–1950) in 1932, which had been the most successful vector of personalism. Though they had initially joined its organized

industrial society becomes a truism. Messrs Sauvy and Fourastié are temporarily grinning and bearing, waiting to take their revenge at the first turn.[10] After a day in the sun of fashion – this conformism of motion, beware of going out of fashion! Since Professor Dumont said everything in 1974 already, the topic will be exhausted. Even if the earth's reserves are too, as it seems ever more.

But 'the ecological problem' is inscribed in facts as much as it is evacuated in the swamp of official discourses. Eventually, we must decide to build water-treatment plants and to create basin agencies to defend water – which they will do by building bodies of water for irrigation and tourism. The pedagogy of a first catastrophe, that of the *Torrey Canyon* (in 1967), had only led to the Polmar plan (rather than the Depolmar plan); and we had to wait until the *Amoco Cadiz* oil slick in 1978 for the French State to decide to take more serious measures. Since it is never done ending, it is the oil crisis which will take over from an ecological movement that is losing its momentum when it comes to raising the energy issue. Despite all difficulties, our economic supertanker is idling. Disregarding the resistances of things and people, projects lead straight to implementation for highways and nuclear plants. Land consolidation expands as agriculture disappears before agrochemicals and agribusiness. The concrete wall progresses along the coasts and around the parks, which are more or less defended. If Concorde and the Rhin-Rhône

---

network of readers in hopes of turning it into a revolutionary movement, Charbonneau and Ellul became disenchanted and broke with it in 1937. Their misgivings were confirmed after the war when *Esprit*, eager for progressive credentials, became little more than a Christian-inspired subset of the left-wing intelligentsia in its successive guises (from fellow-travelling to the New Left), condemning as reactionary nostalgia and pessimistic timidity any critique of industrial society like that of Charbonneau and Ellul – of which Mounier was aware; they are not mentioned by name, but alluded to between the lines in his dismissively titled posthumous collection of lectures on the topic (1946–1948), *La Petite Peur du XXe siècle* (Paris: Éditions du Seuil, 1959), translated by Cynthia Rowland as *Be Not Afraid: A Denunciation of Despair*, with a foreword by Leslie Paul (New York: Sheed and Ward, 1962).
[10]Translator's note : Jean Fourastié (1907–1990) was the prominent economist who coined the term *Les Trente Glorieuses* ('the glorious thirty [years]') to describe France's post-war growth period until the 1973 oil crisis, while Alfred Sauvy (1898–1990), who coined the term 'Third World' for 'underdeveloped' nations, was one of France's first technocratic social scientists and, both in this expert capacity and as a public intellectual, an influential booster of economic modernization and population growth.

canal are struggling, we move on to implementation of the TGV.[11] What else is there to do, if not developing the economy come hell or high water?

And yet, if the sorcerers are powerful, things are deaf, and their propaganda sometimes unleashes uncontrolled forces. Individuals and groups of the ecological avant-garde, previously neutralized, have the opportunity to make themselves heard and to act; but their voice is quickly smothered by that of new converts who are better positioned. And it sometimes happens that ecologists, having made it up the podium, allow themselves to be seduced by the prestige of media or notables. In 1972, the congress of *Nature et Progrès*, at the Palais des Congrès of the Porte Maillot, a symbol of the Pompidou era's lunacy, suddenly draws crowds and media attention. When one has been ignored for so long, how can one resist giving in to the pleasure of being in fashion?

From 1971 to 1977, the ecological movement is on a roll. It then goes after the temple's very foundations: school and medicine with Illich and the *Survivre et vivre* group, created by mathematicians Chevalley and Grothendieck, even goes so far as to criticize an elitist, esoteric science that is contrary to democracy. The movement focuses on nuclear energy (Bugey demonstration), solar energy, Larzac or Lip;[12] to a lesser degree on organic agriculture. But just as important was the proliferation of defence committees that put a brake on technocratic operations that previously used to be developed in a social vacuum. They say there are several thousands of them in France. And sometimes their founders, rather sceptical at first, find themselves in front of a full room of local people, including a good proportion of young people.

Very soon – too soon perhaps – ecology took on a political dimension with the Dumont candidacy during the presidential

---

[11] Translator's note: Acronym for *train à grande vitesse* – high-speed train.
[12] Translator's note: The Larzac plateau in Aveyron was the site of a decade-long non-violent civil disobedience action by farmers resisting plans for the extension of a military base on their lands, finally cancelled upon the election of François Mitterrand as president in 1981. It had become an international meeting ground for all the left-wing protest movements of that era, with something of the atmosphere of a 'French Woodstock', in the words of anti-globalization activist José Bové, who would remain based in a local farm. Workers from the self-managed occupied Lip watch factory in Besançon, another flashpoint of post-1968 social struggles, also came to support the farmers.

elections of May 1974. At the municipal elections of 1977, by going over 5 per cent and sometimes even 10 per cent of the ballots, they seemed to be in the position of arbiters between Right and Left: hence the sudden interest of the big parties, which developed their own ecological commissions. The movement spread to all European countries: in Spain, where ecology is combined with regionalism, a hundred thousand Basques demonstrated against the Lemoniz nuclear power plant. In July 1977, the demonstration against the Malville plant gathered 70,000 people and ecology got its first martyr. But Malville ended in disarray. At the 1978 legislative elections, the percentage fell to 4 per cent of ballots and at the Paris by-election, ecology's star Brice Lalonde only got 3 per cent. And at the 1979 European elections, the ecological movement came very close to 5 per cent. Yet, despite the crisis, the pressure of development still rises ...

CHAPTER THREE

The various constellations of the ecological nebula

The ecological movement is under the sign of differences and contradictions, which is that of life – hence of death: it is not a frozen moon, but a nebula in fusion. Instituted from on high, 'ecology' is also the spontaneous invention of a young age class. A movement, in the full sense of the term, and not a political party – an army; although it has its own minority of militants, variously tempted by the military's efficiency. A social fact, as the trade union movement had been, that takes the most varied forms and whose independent and divergent tendencies no central body has been able to gather until now.

At the centre and on the margins of the maelstrom

In this 'melting pot',[1] we find everything. In it, the *dernier cri* youth that reads *Hara-Kiri* and *Pilote*[2] sometimes runs into the

---

[1] Translator's note: In English in the original.
[2] Translator's note: *Pilote* was a classic comic strip magazine whose content 'grew up' with its initial teenage audience after 1970, when the satirical magazine *Hara-Kiri*, banned for making fun of President Charles De Gaulle's death, was immediately reborn under a new title in his honour: *Charlie Hebdo*.

survivors of the 'earth that does not lie',[3] put back into orbit by the vagaries of history. Marxists discover nature and desire there by way of Freud, Reich and Marcuse. Anarchists, young and old, demand absolute freedom and equality against a progress that has turned totalitarian. This crossroads becomes a meeting point for rebellious teenagers who sense that any profound 'no' to Society must be addressed today to Institutions and Production as it was to the Church in the old days and, in their wake, for a whole mass of opportunists on the lookout for Fashion: young go-getters for whom ecology is the ideal gateway to a career and old fogeys who had previously been turned away. There are the ecologists of a very Parisian *Sauvage*, supervised by the *Nouvel Obs*,[4] and those, more loudmouthed, that *Charlie Hebdo* supervised. And, here and there, some anonymous provincial, attached to his trees or to his river, ready to fight for them because his life would no longer have any joy or meaning if they ever disappeared.

One finds *écolos* of every shape and description, even square. The right-wing or centrist one (who is also a little left-wing) wears his hair rather short ('Mr Barber, please leave my hair longer at the back'). The close-fitting attire and the attaché case are borrowed from the technocrat; besides, technocracy happens to be his area. His reports are well-regarded at the prefecture and they are carefully filed away. For the dose of green (careful, it can turn red for a colour-blind minister) that he puts in his cocktail is scientifically calculated. In commissions, he ends up next to the duly licensed ecologist who somehow manages to put up with his leftist assistants. This one remains pent up in his dungeon: his lab and his science; if he ever gets out, it is to manage the nature reserve entrusted to him by Esso Standard. He is a scientific ecologist, he will never compromise with thousandths, and he can be entrusted with impact studies, sources of credits for his institute, if not other

---

[3]Translator's note: '*La terre ne ment pas*', from a speech by Marshal Philippe Pétain, was a slogan of his wartime collaborationist Vichy regime, whose National Revolution had ruralist pretentions.
[4]Translator's note: *Le Sauvage* was France's first magazine of political ecology. Born under the wing of *Le Nouvel Observateur* as an outgrowth of its ecology column in 1973, it did not long survive its founder Alain Hervé's attempt to make it independent by buying it back from the senior mainstream magazine that tried to reannex it as a mere supplement in 1979.

bills. On the right wing also and sometimes at its extreme, joining the extremists of the other side, we find certain paratroopers of naturist commandos. They too have short hair, even though they expose it to the sun on principle, and they do not trifle with reason and morals; but since some of them push fundamentalism too far, like Mgr Lefebvre, they are suspect in Rome, and end up being mixed up despite themselves with the bearded types whose dubious hair brushes them the wrong way.

The bulk of the species, both male and female, can be spotted from afar thanks to their flowing hair – we may speculate that the addiction of previous generations to a Roman haircut caused it to grow back with a vengeance. Short hair, long hair, who is next? A sign of the military under Chilperic,[5] and then of the artist under Fallières,[6] why did it become that of the non-violent? In any case, there too, the nature of the uniform and the length of the hair are determined by regulations. However, could it be that this attire that helps us distinguish friend and foe amid the fray is on its way out in the cities?

But the thickness of the foliage should not hide a forest composed of very different varieties. The list is endless. In the chorus line, the gentle, hash-scented hippie returning from Kathmandu as a convert to Buddha appears next to the bespectacled eco-Trotskyite militant, rationalist and abrupt, and the Breton country squire driven mad by the DDA's bulldozers.[7] And there are so many more ... Among the mass of leftist high school students, we should always expect the appearance of an old notary who has been expropriated by the ZAC.[8] Only the prefect and the Communist Party's ecologist, both kept away by their mother's illness, had to be excused. The originality and richness of the ecological nebula consists in the bringing together of social and personal differences. Its richness and its strength, but also its weakness.

Despite being co-opted by the authorities, the ecological nebula is bound together by a feature common to the full range of its

---

[5]Translator's note: Merovingian king.
[6]Translator's note: Belle Époque president.
[7]Translator's note: The erstwhile *Direction départementale de l'agriculture* would sometimes carry out public engineering mandates at the request of municipalities.
[8]Translator's note: *Zone d'aménagement concerté*, an area singled out to be provided with public infrastructure in view of private development.

components: marginality. It brings together all those whom, for all kinds of reasons, industrial society has pushed to its borders; and since it makes just about everything its business, that means almost everybody. Conservative or revolutionary, teacher or student, peasant or craftsman, metallurgist or even blacksmith, we are all feeling more or less threatened by this disorderly order from which we draw wages or profits. And we no longer know whether we should defend it or bring it down. Like our interests, our values feel threatened: the societies where life is easy are the ones that are ill at ease. And the most demanding or the least strong give in to anxiety. How can we raise our gaze to face the tidal wave, the dizzying and poisonous mushroom that looms over us from now on? Since our legs are not swift enough, we might as well flee into some dream or entertaining agitation. Neurosis, just as much as revolution, is an engine of the ecological movement: there is something of the 'ship of fools' about it – and there are those who are fools for being too wise. Let us not look down on neurosis too much, for if revolution comes from the head, neurosis comes from the guts that know a whole lot more. Revolution is too often inspired by a narrow-minded reason that produces ideologies and barracks, whereas neurotic rebellion spontaneously arises out of the depths of suffering. It knows too much, which is why all it does is shout. And reason has to translate the signs it directs to us.

The ecological movement is home to quite a few neurotics great and small – without forgetting those who are just acting the part. As always, on this dung heap (fertile to the extent that it stinks, we should keep in mind), religion prospers. All the wilder as it must trade in its cassock for some jeans and must put on the mask of science or politics. In this green Garden of Delights, gurus who prophesy a mystically scientific revelation prowl for disciples, some wonderworkers offer their drugs or foods of eternal life and others various aphrodisiacs to give back to desire the potency to lift the universe. Since religion is the last freely available good, beyond the control of academia, trade unions and parties, this hole opens quite a niche to fill in. Something which a few very realistic, slightly crooked minds are busy doing, taking it upon themselves to provide the people with drugs, truths and natural products. Besides, this foam that the maelstrom rejects on its margins can be recycled in the guise of a host of cultural, artistic, tourist or advertising gadgets.

But far more marginal is the serious ecologist – be he intellectual or militant – for whom ecology is neither merchandise nor career nor entertainment. It is up to him to clean up.

This marginal character is shown by the place taken by sects. It makes sense: when the Church rules, celebrating high mass, it drives back those who refuse it to its periphery, into coteries of devotees. Ecological sects can be related to the Protestant ones from which they sometimes derive: they all insist on the purity of orthodoxy – but each one has its own. They live in their ghetto, apart from the bulk of the ecological population that they join on important occasions. They have their antipopes, their holy books, their rites and their economic base. Their principle is a more or less strict naturism or vegetarianism and, to celebrate mass, they strip like others dress up. Unbeknown to them, their cult of a virgin, unpolluted nature is similar to the hygiene practised by the official Church. At *La Vie Claire*, natural products are sold by maidens in white blouses like those at the hospital. Naturist sects are wary of the temptations of the flesh and of desire which are however celebrated by other ecologists. But one extreme discreetly leads to another. The faithful mortify themselves by refusing the aphrodisiac poisons of alcohol, tobacco, red meat (white or rosy is better). They come from the cold, that is from the north, the land of *Reformhäuser* where, as early as 1961, one could find untreated fruits at the Hamburg market. Some fundamentalists go so far as to consume grass, and even, in case of illness, earth, which is a panacea. They have refused mainstream medicine, its chemistry and its vaccines. At a time when the countryside is registered with social security, they are retrieving the tradition of healers; or, at the very least, they practice herbal medicine, homeopathic medicine and acupuncture.

Sects are at once gnostic and argumentative. There too, Revelation is provided by Science. But this science, bioenergetic, (again bio) dynamic, macro- (or micro-) biotic, parapsychological and so on is persecuted by academe. This specifically Western avant-garde is attracted to Eastern mysticism, as are many intellectuals and young Anglo-Saxons. It takes up yoga and Zen, delights in the play of yin and yang, which is less popular in the countryside of the Catholic Midi.

However, let us not speak ill of sects; despite their foibles, they have their virtues, and they have sometimes succeeded where other ecologists have failed, by founding communities. Their truths may be stereotyped, but they believe in them and keep to them. They may wallow in their ghetto, but they manage to build it and live in it. Their morals may be somewhat hypocritical, but among themselves, their zealots practice it. They practice their virtues of mutual help, work and thrift, without which nothing lasting gets done. Good or bad, they have created societies that have preserved a share of originality despite the State's stranglehold on social life. They therefore show the way for other communities and the pitfalls that are best avoided.

Where Nature switches from Right to Left

But the great novelty about this conservationist, if not conservative movement, is that it takes on a more or less revolutionary left-wing tone. Which is understandable, the *new look*[9] bourgeoisie having played the card of technical progress and the liquidation of agropastoral societies, just as the old one had purported to simultaneously carry out industrialization and the conservation of the countryside. A switch had occurred between the Right and the Left as there had been one between war and peace. The defence of the 'earth that does not lie' has gone from Pétainism to the extreme Left, while the more status quo parties like the RPR[10] and the Communist Party persisted in the defence of movement. And we see conservative academics, reacting against their students, take a stand against ecology in favour of a development whose certain effect is the upheaval of the natural and social order. The man of the Right, like a good Catholic, is not attached to values or an immutable orthodoxy, but to any given status quo: challenging the expropriators of the Ministry of Equipment or the Larzac is an attack on Order, just like threatening property in the past. In many areas, there is a complete reversal. What was unsayable on the Left at the time of Keynes and Stalin suddenly goes without saying.

---

[9]Translator's note: In English in the original.
[10]Translator's note: The Rassemblement pour la République was the heir to the Gaullist movement.

Ecological leftism, and to some extent the Left *tout court* on its heels (at least some elements of the CFDT,[11] the PSU and even the Socialist Party), is opposed to nuclear progress, rediscovers the countryside, rootedness and local societies. It breaks with the centralized *Gleichschaltung*,[12] both nationalist and internationalist, of which Communist orthopraxy is the perfect model. While the Right, true to its old habit, unhesitatingly continues to sacrifice its values to the ones in its wallet.

Between the Right and the Left, this switch about progress is explainable. By criticizing development (let us not say progress anymore), leftism retrieves the libertarian tradition of a simple and natural life according to Jean-Jacques and Babeuf. At its Rousseauist origin, the naturist revolt does not fit into the polemical framework of Right and Left, for it is both ultrareactionary and ultraprogressive. In principle, nature is on the Right. Ecology reminds us that man is not a god but a finite being, dependent on his environment and his history: the progressive dreams of the future while the naturist dreams of origins. Whereas the Left insists on the universal, ecology emphasizes specific differences. Until Illich, the critique of science and schooling was right-wing, but Illich is only taking up again that of the prophet of 1793 and the *Discourse on Inequality*. Rousseauism is consistent. Faith in the goodness of man entails that it also be granted to the matrix in which he was formed. And if nature is right, civilization is wrong, its progress is only that of constraint and hypocrisy. The laws that nature imposes on freedom are reasonable and have nothing tyrannical about them, and it is not in the city that the latter can unfold without fear of the police and of red lights. Reason, morals, that means being natural, naked, like an animal, a child or the noble savage. Clearly, progress is on the Right!

The left-wing ecologist can be more ecologist than leftist, or more leftist than ecologist; but in this case, if the Party chooses the highway, he will prefer to lie low. Chances are he will join it again pretty fast so as to ensure ecology's 'presence': the formula

---

[11]Translator's note: The formerly Catholic trade union confederation behind the Lip factory takeover.
[12]Translator's note: A German term for the forcible imposition of party control and the party line throughout all institutions of a society, drawn from Nazi parlance and practice.

also applies to the UDF[13] or EDF,[14] its application being trickier for the Communist Party. Ecology's relations with the PSU are simpler, as it is a less demanding party due to its failure to be a party.[15] This at least allows ecologist sympathizers to decree in an earnest tone: 'Everything is political.' Which is not untrue in the long run and makes you look like an expert in front of the rookies from the Muséum national d'histoire naturelle.

Part of ecological youth has its roots in Christianity, though it is wary of acknowledging that fact. Jeans may take the place of the cassock and slang that of Church Latin, but too systematic an anti-Christian stance is reminiscent of the anticlericalism of some former clericals. They still feel the need to take a distance. Though rebelling against the churches, they are still motivated by faith and the evangelical values these betray, and we may find among them a good many Protestants or Protestantized Catholics.

Hence non-violence, conscientious objection, which take up anew, after the break of war and Resistance, the tradition of integral pacifism. Why had this tradition been broken? It is the question to ponder if we do not want history to repeat itself. We rightly oppose the extension of the Larzac camp, but we dream of going underground and sympathize with liberation wars led militarily, advised and equipped by the other side's staffs. Christian moralism reappears in the rather austere feasts at which organic fruit juices flow freely; in the critique, ever so justified, of a wasteful society that calls itself affluent. But it would be political to ask it about the nature of this plenty. It is also the Gospel that is the source of the positive valuation of poverty, of everything on earth that is oppressed and suffers: the excluded, prisoners, the mad. The only hope against development rests with the underdeveloped: the hungry of the Sahel or of Brazil's Nordeste; Vietnam, once heroic, and China, once a People's Republic, have lost some of their prestige. As for India, it does not have enough political existence. We come here to the other pillar of Christianity, justice, rather ill-fitted to the pillar of freedom,

---

[13]Translator's note: The Union pour la démocratie française was a centrist party created to support President Valéry Giscard d'Estaing.
[14]Translator's note: The State utility Électricité de France.
[15]Translator's note: The Parti socialiste unifié was a compound of left-wing splinter groups and dissidents. Committed to self-management, it supported the May 1968 student uprising and the Lip factory takeover.

which makes that edifice a trifle wobbly. And it is freedom (that of the children of God or of the Holy Spirit) that we will now find in the area (I mean at the level) of ecology's libertarian dimension. Having escaped from the sacristy and secular school, and then lost its virginity at *Charlie Hebdo*, Freedom is a pretty raggedy girl, albeit a warm and friendly one, who opens all kinds of horizons until now hidden by nature. One fine day in May, she ordered the schools to 'take your desires for reality', an inadvisable formula for anyone who wants to make a reality of his desire – though which one exactly? Anarchism, with a keener nose than Marxist dogma, has located enemy number one, whose liquidation had been postponed until then: it attacks Power, which so far does not seem much affected. It denounces all forms of repression, calls for the elimination of police and jails; against economic and political centralism, it demands a democratic, self-managed society, relying on an autarchic base and simple, light technology. Respectful of differences, anarchistic ecology moves closer to various regionalisms – but beware of the nationalist temptation! Freedom being indivisible, the critique of the State becomes that of Society: after the barracks, it attacks the family, and after Capital, Labour, which was sacrosanct until then.

All freedom being due to all, there is no authority or power that is not being challenged. We are wary of leaders and stars, even of elected officials. We refuse to put up with the truths of the Army, School and Medicine (with more hesitation when it comes to science). For starters, we defend *birth control*[16] against the population explosion (not without some uneasiness about the Third World whose opinion is taboo), but we disagree about the pill, this product of chemistry. We support the MLF[17] that demands the right to abortion, and in the name of the liberation of desire according to Reich, that of homosexuals and finally of paedophiles and so on. For no constraint or rule imposed on sex is forgotten, not even the one that a shared love imposes on itself. These antimorals, being too systematic, turn into a religion of sex, as the orgasm gives you revelations and powers once granted by its refusal. In certain groups, sexual freedom too becomes a constraining rule that does

---

[16]Translator's note: In English in the original.
[17]Translator's note: Mouvement de libération des femmes ('Women's Liberation Movement').

violence to nature. And it is the desire of others that we end up taking for reality.

But as we have seen, this trend is but one among others. And the ones we have just briefly mentioned mesh, not just in the movement as a whole, but also within each one of its members.

Communal microcosms and silent majority

Until then, the issue had been not to defend or change life but the regime. For the Right, it went without saying that it was good, and being busy producing was all there was to do; for the Left, you only had to wait for the revolution, the social and the cultural being hitched to the Communist locomotive. But after de-Stalinization and its avatars, in spite of the boost of Maoism and heroic Vietnam, depoliticization had slowly won out over the post-war era's politicization. And post–May 1968 disappointments had only heightened it. Nothing had come out of the regime change in the USSR and everywhere else, except other schools, other armies, other jails. Better tomorrows indefinitely promise better tomorrows; and ever-impatient youth wanted to change life that very day. The centre had betrayed its hopes; it withdrew into the provincial margins the tractors had evacuated.

Far from the city, they wanted to move into practice and create another society, on a human scale, poorer but more natural and freer. But if there is a place where reality bumps against desire, it has to be the site of the silence and the demanding labour of the fields, the site of the house, isolated or in a village, communal or family-owned, where every day you find yourself among men and women, parents and children. Communities quickly learned to distinguish the dream from the reality, play from seriousness, the possible from the impossible. This is why so few managed to last.

Since May, in France, following the United States, the lifeworld has been discovered: the sensible realities near at hand of which daily life is made, both public and private. In other words, we are getting out of political alienation that leads astray into myth or ideology for the benefit of monopolistic or competing firms devoted to the religion of power. But this putting back in its place – not necessarily a refusal – of politics takes all kinds of forms. In the ecological movement, communal utopia goes hand in hand with its opposite: the sometimes-blinkered defence by countless committees of whatever happens to exist here and now. It is not just a few bearded fellows who turn into ecologists, but any Frenchman – this one however an ecologist who does not know he is one – when the flood comes to lap over his doorstep. What peaceable angler fails to ask himself that question the day he comes across a river that has been turned into a sewer, where the fish flash by, belly up? What homemaker does not wonder about the ocean's future when, on a beach festooned with plastic, glass and fuel oil instead of algae and shells, she must watch the spot where her kids put their feet? And what can you say when your boss tells you that you are going to have to leave Salies[18] for Pampérigouste[19]-upon-Yangtze, where factories have just opened to ensure employment? And, worse still, when you find gentlemen with a surveying rod driving red stakes into your orchard? Your environment is like your body, if it is healthy you do not feel it; it is only when all kinds of interventions assail it that you become aware of it and talk about it.

Thus, in the United States, and then in Europe where there is a lot less space to go around, people had to come to the realization that not only the quality of life, but life itself – which means first mine and yours – was threatened by the gravel pit, the power plant, the military or leisure camp and so on. The time when the village attracted the thunderbolt, that is the highway, is over: we are not against it – provided it is at the neighbour's place. Hence the ping-pong between notables, refereed by the Equipment Ministry. And the proliferation of defence committees of (hence against) this or that. The way things are going, there will soon be no polluter who, to defend his environment, will not belong to some defence committee against the pollution ... of others.

What analysis differentiates and classifies, reality mixes and opposes in disorder. Hence the hazy, sparkling look, with the occasional lightning flash, of the ecological nebula. Wanting to seize everything to reply to the opponent, it harbours within itself

---

[18]Translator's note: Salies-de-Béarn is the town closest to the hamlet of St-Pé-de-Leren where Bernard Charbonneau had his summer home in an old barn.
[19]Translator's note: A fictional Provençal town in the stories of Alphonse Daudet (1840–1897), Pampérigouste has come to refer in France to any location in the middle of nowhere.

all kinds of contradictions that are on the minds of ecologists. Between such diversified tendencies, there exist perforce tensions and conflicts. They come back together about Larzac, nukes, Lip or Chile, but that agreement remains fragile. The anarchists and classical Marxists for whom ecology is merely an add-on have trouble getting along with those for whom it is of the essence. Those who worry about the degradation of their countryside and their food are wary of believers and of traders who sell more or less 'organic' products; carnivores are opposed to vegetarians, and those who remain hunters to the anti-hunt crowd. Naturist progressives for whom science remains taboo denounce as reactionaries and prefascists those who condemn it, and the more or less integral naturists whose background is in scientific ecology castigate those who allow for human thought and action. Those in favour of 'changing life' argue with those who favour 'regime change', as mystics do with rationalists and so on. These contradictions and conflicts could be displayed in the form of a chart:

Violence/non-violence
Electioneering/election boycott
Science, reason/subjectivity, dream
Asceticism, puritanism/desire, festival
Organization/spontaneism
Natural determinism/human freedom

And the list is still open.

Furthermore, in a movement of revolt that brings together discontents and marginals, personalities are strong and numerous, and the refusal of leaders is at times hard to distinguish from the desire to be one. Hence the multiplication of orthodoxies and splinter groups that excommunicate each other as in the heyday of surrealism and situationism (*viz*. the internal quarrels of *Survivre et vivre* and *La Gueule ouverte*). That which separates sometimes causes people to forget what unites. Will the ecological movement be able to take on a modicum of shape? If so, what will that form be? And if it achieves the needful unity, is there not a risk that it will lose in richness and spontaneity what it will gain in efficiency? If it does not want to fail, be repressed or get co-opted, it will first have to confront its own contradictions in all foreseeable cases.

To remain on its guard, to practice examining its own conscience, that is, to self-critique (we can no longer talk of autocritique, since lies have devalued it so much).[20] This is what this author's book is attempting as a participant in the movement, and based on his values, but one is then less likely to please one's friends than by supplying them with received ideas or miracle drugs. And this critique will need to be carried through down to the very principle in the following chapter.

Will the ecological nebula cool down? By superficially losing its solar heat, will it become a planet, less hot to be sure, but one where clouds will condense into springs, where trees may take root and people get a foothold?

---

[20]Translator's note: Charbonneau is here distancing himself from the Communist practice of autocritique.

PART TWO

Roots: Foundations of the ecological movement

Let us now consider a tree's roots. How difficult it is to expose them to the light of day! How fine they are to be able to worm themselves deep into the earth in search of its juices! But how powerful they are to anchor themselves into the rock and hoist their way up there to the trembling treetop that defies lightning.

CHAPTER FOUR

Nature or freedom?

Distilled to its essentials, the ecological movement comes down to two watchwords discredited by overuse: nature and freedom. Nothing less than the carnal and spiritual aspects of the human world – or in other words, of the world altogether. Since Rousseau, individual freedom has been associated with the love of nature. And we have seen to what extent feeling for nature, more common in the industrial countries of the North than in the agricultural ones of the South, prevalent among the rich intellectual bourgeoisie before it reached common people, goes hand in hand with the rise of development and individualism. If we review the criticisms and claims of ecologists, they roughly fall under the two headings of nature and freedom. To the first belong the protection of the environment, that of endangered species, the struggle against pollution, against excessive land development, for green spaces, against the threat of a nuclear disaster and so on. To the second belong calls for self-management, for women's liberation and sexual freedoms, anti-militarism, anti-statism, regionalism, attacks on the police rule of nuclear power and so on. The reader may want to indulge in rounding out this classification, formal as it is like any other. The point is now to clarify the relationship between nature and freedom.

It is no accident that these two keywords give positive definition to the ecological struggle. For nature and freedom negatively sum up the endeavour of a society which, to go all the way in destroying the earth, must wholly encompass all its inhabitants. If these two notions illuminate the ascending path the 'ecological movement' must take, they also shed light on the descending path industrial civilization is hurtling into. The latter may justify the pillage of nature in the name of human freedom, but it is actually annihilating the one with the other. The devastation of land and sea goes hand in hand with the chaotic proliferation of techniques of social control applied to peoples as well as to individuals: as much as to master natural complexity, the computer will be used to control human spontaneity and diversity. At a time when individuals, caught up in accelerating change, lose their memory, an implacable Memory, like God, will recapitulate everything. If this fails, a card will be perforated with one more hole. Let us keep to this example; here too the reader may have fun classifying the costs of development under the different columns of nature and freedom.

'But what nature? Which freedom?' lovers of definitions will ask. The nature and the freedom they have in front of their eyes and in their heads. If they do not yet know what they are, I cannot tell them. Nonetheless, I will try, working my way against centuries of confusion and hypocrisy surrounding those polluted words, up to the crystal clarity of their origins.

Nature

What is nature? An idle question for the pioneer who is struggling against it; he has other things to do. And for the suburbanite who dreams of desert islands. A question almost as futile as 'What is freedom?' for one who has already lost it. Nature? You discover it the day you come out of it: it was distinguished from culture when the sky that was once held up by the world ash tree Yggdrasil became enclosed within the sphere of steel, bricks and smoke of Krupp's Germany. Even if we do not know how to say what nature is, we know what it is not: the discarnate order of human laws imposed by the iron rule of barracks and police stations, the sharp-toothed machinery that tears down oak trees and tears open hillsides, the grey concrete glacier that covers them, the stinking cloud that rots the air and water. In a word, human artifice.

This nature that I discover within my grasp as anybody does is not the old concept of Greek philosophers or medieval theologians, nor that of science, indefinitely definable by mathematical signs and vanishing to infinity. It is not the cosmos, it is not the universal matter to which the infinite diversity of appearances can be reduced. In the evening, on summer nights, at the bottom of the black void, we see it shining, and our spirit rises in vain towards those up to now inaccessible lights. From above or from below, in the macrocosm or the microcosm, that nature eludes our all-too-human power and knowledge. It is the whole, whereas we are almost nothing. There is not a nebula, not an atom of our flesh that it does not constitute; and there is nothing we can change to its laws. All-powerful and mute, it puts up with us for a little while; and should we show contempt for its laws, which it did not even bother dictating to us, with a breath it will blot us out from the universe. To protect nature? What presumption! It is forever invincible and its sanctions will be terrible. We can pollute the ocean and thereby destroy ourselves; for a long time afterwards, a surf heavy with hydrocarbons will beat on dead beaches. It is not the protection of nature that is at stake, but the protection of man, by and against himself.

Nature, that of human beings and not that of gods, is not this physical–metaphysical universe, but this planetary accident: the earth, of which I am made and that feeds me, whose breath is air, whose blood is water, without which life would not exist. The nature that I love is alive, it is this unique space-time unknown to mute and petrified other worlds, where the sky is a whisper or a rage, the land or the sea a sudden flight animated by the multitude of flora and fauna in their ever-shifting forms. The earth is the life-planet, green and blue, without equal for a human gaze or the human mind whose giant body it is, minuscule though it be in the universe, a body as precarious as our own – as we know since Hiroshima. It is this body that nourishes with all the juices of the earth this accident within a planetary accident: man, and more precarious still, the human being I happen to be.

Man is nature, as we are getting to learn, now that we have come out of it for good and that our means threaten to destroy the earth, and us along with it. And nature is manifest in this man: in my individual, personal yet impersonal body, subject to the uncertain laws of matter and of life, which, distinct from my consciousness, is nevertheless its root. Nature is everywhere. All around me, in the air that I breathe and the rock under my feet; I dare not say my 'environment', since this fashionable term suggests that we do not really know what it is and that it is exterior to me.

Nature is my body, sensitive and active, without which the spirit would remain an abstract idea, and through which everything is spoken: the lightning and the haze. Whose limbs, woven from sinews, allow me to survey and seize the earth, to advance in its changing space as one progresses through time. And it is present even in my brain, where the precise point at which matter becomes spirit will no doubt never be found. Thanks to which, one leading and animating the other, I can yield to the human urge, at once legitimate and usurped, to create a work and to fashion the soil as I wish; this is also called wielding my power, a passion that can degenerate into madness. But this body, this ally to which I owe everything, will nevertheless betray me. It is already busy doing so.

If my freedom is not just a word, if I experience it in pain and in pleasure, in solitude and even in love, then it is here, in this place where all the powers of the universe interact and which, in spite of this, remains the most particular one of all: my individual body. At the same time as it makes me a real person, it forces me to merge back into that nature from which I feel distinct. As much as my spirit, it is this body that makes me unique by situating me in space and time. Without its particularities, what would I be? Without the sound of my voice, the colour of my hair, the style of my gestures, the reactions of my temper, which, even if I had no name, make me recognizable, who would I be? What would my freedom be? Not even a shadow amid endless night!

Freedom

The other – and thus the reverse – pole of the ecological movement is freedom. Once again, not the discarnate freedom of the metaphysicians or the freedom, identified with the State, that holds up its barracks and its prisons. The freedom which ecologists sense, spiritual and carnal, belongs to all, because it is in everyone. This being said, there can be no question of providing it with a precise definition, like that of what it is not: the constraints inflicted by nature or by man. But if it cannot capture it in a concept, instinct shows us which way to go to find it, and the ecological movement has rightly seen it. Freedom – what a scandal! – is the claim of the part against the whole: of the nation within the empire, of the homeland within the nation, of the village within the homeland and, within the village, of the individual. The individual is the one who is the irreducible fundamental without which freedom makes no sense. In proclaiming the individual's freedom, the great Revolution was right. In relation to this one, all other freedoms are more or less abstract conditions for it, which lend themselves to justifying its negation. And having first turned our back to the universe, the more we go towards freedom, from the State to the town, from the town to the family, from the family to each of its members, the more the idea of freedom gets fleshed out, until it finally delights or bleeds in this ultimate existing being: one's self. There is no focal point more intense. For if animals are dimly aware of their death and of their life along with it, human beings become conscious of it. To see, to feel, to live and to know it: such a one lives twice. And nobody can do it in his place on earth – still less on other planets made of ashes or fire.

There is but one man: a lone one. This is why he goes towards his fellow, who is one too. The first thing he asks of him is to be himself and to give of himself freely, for everything that is obtained by force is but a violation, a lie and an ersatz. In a world, in a society that is nothing but constraints weighing on the individual, the human chokes. He will prefer to live in a poverty of his own than in imposed wealth.

To be oneself, to live and think by oneself, without doubting one's identity nor the other's, such is, as much as bread and water, the fundamental need and the originality of our species that manufactures individuals – even if they hardly are. The one that I am – that we are. This is any human, both singular and plural. The concrete universal of the old philosophers? Behold him – behold her – an individual and a freedom, at once masculine and feminine.

But although man *came out* of nature, he no longer quite belongs to it; his dream has torn him from his mother's womb. Born of the universe, he dreams of another world in which absurdity, coercion and violence, suffering, old age and death – in short what for lack of a better word may be called evil – would be abolished. By dreaming of non-violence and of freedom for all, of a paradisiacal earth and society where everyone's love and desire would replace the laws of nature and society, the ecological movement obeys in its turn the invincible drive of the human spirit. 'Take your wishes for reality!' Nature on its own would never have given such counsel.

Man is nature, but he is also supernature. Some demon drives him to constantly break out of the circle that the universe draws around him: to say no to the natural and social given if it is not responsive to his demands – and they are insatiable. Man cannot overcome this rage for progress that makes him bang his head against the walls and, when he can escape, look further for others he will persist in knocking down: always chasing Knowledge beyond knowledge, Justice beyond justice, Freedom beyond freedom. Ceaselessly seeking the breach that will allow him to get out of his prison in order to build his house. Man's freedom? 'To the creation that surrounds it, it says no. To the love that seduces it: no. To the voice of the People, to the order of the Prince, to the cries of the multitude: no, no, no. It is the eternal naysayer that comes in the way of every thought and every existence; the untamable insurgent that only trusts itself, that has respect and consideration for none but itself, that only respects the idea of God to the extent that it recognizes or not in God its own antithesis, always itself.' Proudhon has accurately described the passion for freedom. How could nature, any more than society, long tolerate such a demon? But Proudhon knew well that freedom goes so far as to be its own master.

The contradiction between nature and freedom

Nature constitutes me. Yet I cannot wholly identify with this power that bestows upon me, along with life, decay, suffering and death. And I am not the only one facing this condition. Nature is not the divine Whole in which the individual agrees to dissolve; if this Whole is all there is, I fail to see why the part would purport to add itself to it. A child of the earth, I remain a stranger on it, as it is the Other to me, and the power of the feeling that binds me to nature depends on this very difference.

Nature has given me life, but I cannot find life's meaning within nature – only the weight of necessity or the swirl of chance. Its law is that of entropy, even if life can claim local victories here and there, as does freedom over the course of life. And it is not within my power to accept this iron law as plants and beasts do; my conscience must inflict on my mind what my senses inflict on my flesh. How is it that, sooner or later, even if I flee it, nature inflicts upon me the ultimate suffering: thought? Perhaps meaning consists in this very nonsense that forces humans to act upon the universe in order to give it meaning. I know very well that this is doing violence to nature, with all the risks that this implies. Here too, how do we make sure that this other power of life, human freedom, does not turn into a power of death that finally destroys when it wanted to create?

For human freedom conflicts with nature, except on one point: life, which nature grants, on the condition that it follows the game's strict rules. The one would want to do everything; the other teaches freedom that you cannot do everything in this fragile sphere – this ecological niche – upon which life depends. If man has managed over time to escape from his familiar prison, it is by enclosing himself in a steel diving suit that feels a whole lot tighter. Freedom is imagination, invention, and nature teaches us our limits: what our consciousness must first take into account. By denouncing the excesses of progress and of development at any cost, the ecological movement, born as it is of the libertarian explosion of May 1968, reminds us that, for humans living on earth, everything is not possible.

But if there is *one* order of nature that we can to some extent escape only by obeying it, its reasons are not exactly those of human reason. The latter keeps trying to hone its French-style garden out of the inexhaustible, magnificent jungle. Yet how stiff and pitiful this reason is, compared to the ever-richer reason that has allowed life to blossom on earth, and even the universe to arise! Every time we think we have finally understood it, a new door opens through which it escapes the reach of our science. It is in vain that human reason seeks to decipher nature's reasons in order to become its master and – who knows? – perhaps, one day, its creator. The more we go on, the more the order it unveils becomes icy and empty, and it seems we are rising towards the very throne of nothingness. And if nothing else comes to enlighten man, the few secrets that nature agrees to reveal to him may only lead him to his ruin.

This supreme good drawn from universal non-being – to exist and, in so doing, to allow the universe to exist in me and around myself – nature grants to me at the price of its exact opposite. And even if my philosophy accepts death in theory, my instinct refuses it in practice. Especially when it ceases to be Death and becomes a particular person's death, mine or, even worse, the death of someone I love, since that one I will experience to the bitter end. What ideologist would bless the knife put at his throat or the agony of a loved one? And we should not forget the revolting refinements that precede it, which medicine only avoids by prolonging them and by finally leaving but one way of putting an end to it: killing. In nature, death is the price of life, but this soothing formula is cold comfort when death comes. There is nothing more profoundly human and natural than struggling to escape it; beasts are no different than humans in this respect for, in order to avoid getting killed, the fight-or-flight reflex will take over with any chance they get. Plants are better at bearing death. But the more we ascend along the chain of beings, the fiercer the fight: evolutionary progress is the progress of agony.

The law of nature is the submission and sacrifice of the individual to the species, of the species to the genus, and of the genus to … let us say evolution; this way, the end becomes clear and positive. Nature multiplies individuals and sends them forward like a general does with his soldiers in the attack, for reasons that the rank and file cannot comprehend. And as for this general, whose name is not known to me, I also do not know where his headquarters might be, but he has no reason to worry: from the remains of individuals and of worlds, he never stops allowing new armies to come forth. Hence nature's imperative as proclaimed until now by gods and leaders: 'Increase and multiply!' Something each one of us can feel free to translate as: 'Diminish and disappear.' This is a divine law that men and women are only just starting to question. And it may be that the human species is itself but an individual of sorts, sent forth toward the goal by the Unknown Field Marshal.

It is this multitude rushing forward – no backsliding is allowed – that is pruned and selected by the struggle for life. Nature is as unaware of the rights of the poor and the weak as it is of those of individuals. '*Vae victis*'[1] is its byword; it treats the vanquished as fodder for the strong, regardless of whether victory has been secured by force or through cunning, that is, by deceit. However, generally

---

[1] Translator's note: Latin saying meaning 'Woe to the vanquished'.

speaking, the carnivore kills according to its law and its needs; once its appetite is sated, a peace of sorts, an ecological equilibrium sets in, ensuring the harmony and tranquillity of the ecosystem.

The feline usually kills for food, and should it occasionally kill for pleasure (as the cat does with the mouse), the ecologist will tell me that it is practising. In any case, that sadism remains innocent. It is not the same with the predator par excellence, man, who plans in advance and organizes the death of everything that does not belong to his species, or even to his group. But in order to say no to this added level of the struggle for life, the war waged by a hydra with millions of heads, it is no doubt necessary to escape the grip of that second nature inscribed within the first: society.

That is why, born of the earth, I dream of another world that would not negate my desire and my reasons. Where life would not be paid for by its decomposition and termination, where I would not have to endure sooner or later the suffering and death of those like me. Even if it is folly, I can only rebel against this order where the strong devour the weak, where the universe, the species, society and its armoured twin, the State, grow on a manure of sacrificed individuals. And I maintain that the shiver of a lamb being slaughtered is a more trenchant argument than the sharpest reasonings and knives.

But when I purport to put some order in this disorder, I often make it worse. For this earth is still mine: my origin and my element, where I am 'like a fish in water', the home I inhabit and whose fearsome wonders I am far from finished taking stock of. The human mind dreams of peace and unity, and it discovers otherness and contradiction – no doubt in order to resolve it. To live in order to overcome this contradiction among many others, such is the freedom of the spirit: my very being. But the five letters arranged into this word 'being' hardly suffice to say how sharp is the blade that cuts to your heart.

To an extent that one would be hard-pressed to set a limit to, man and nature are alien to each other. Man is not nature *or* supernature, nature *or* culture, but both the one *and* the other. And each time man allows himself to be reduced to just one of these terms, he negates himself and turns to stone, so to speak. As we are now finding out again after having lost sight of it for a moment, nature is this warm, heavy body that we cannot deny without destroying ourselves: a necessary condition, one would have said in other times. But it is not a sufficient condition, for man, rooted as he is in the earth, also dreams of freedom. The life he draws from the depths of the soil is what makes him stand erect like a tree; and up there, his thoughts, like leaves, dance in the light and wind. If these thoughts exist, they will in turn give the roots the strength to reach still deeper, until they graze against the solid rock of ultimate foundations.

CHAPTER FIVE

Nature and freedom

Nature or freedom? Nature *and* freedom, the main thing being the conjunction which both distinguishes and unites the two terms.

But two different truths at once, that is too much for our little mind: their contradiction, and even their tension, is unbearable. We must have the one or the other, and so we have to confuse them by reducing man to nature or nature to man, like the naturists or progressives we all are to some degree. Such a nature or such a progress are but polemical half-truths between which debate can rage on forever, each side being half-right. And being thus denied, the contradiction between nature and human freedom grows unchecked. By contrast, the one who bears that contradiction will discover that nature and freedom, because they are distinct, are united on earth, around him and within himself, as well as in history.

Nature and man united in the human environment: Town and country

Man is the fruit of nature, which remains virginal only insofar as he is excluded from it, for he does not inhabit the jungle like the great apes, and the very fact of standing on his feet and using his hands, and all the more so his brains, compels him to interfere. He must put together a shelter and feed himself, always getting better at it; there lies the risk, if his freedom is not strong enough to remind him of his limits. He feels lost in the deserts of ice or burning heat where he seeks an oasis, and he feels choked in the jungle where he opens an ever-wider clearing, until the forest is reduced to a strip – and then a dot – of darkness, over there, at the horizon of the fields: what he calls a wood, where he likes to walk.

As for pure nature – the open sea or infinite space, we venture in it only on some carefully designed vessel. The nature where we feel comfortable has been arranged by human work (but not covered with concrete by developers). Except for some areas of the coast and of high mountains, our environment is the fruit of a conquest and of an accord between man and his surroundings. Europe means the city whose dramatic expansion is driven by cars and the eruption of high-rise buildings; but even more so, it means the countryside, which in France still covers over 90 per cent of the land area. What we mistakenly see as a steppe is actually a countryside of invisibly delimited open fields, and what we take for forests is old trees and woodlands carefully trimmed by cattle's teeth and the peasant's scythe. Nature is but the material of the landscape, which is an edifice of grass and pruned wood or even of assembled stones, like those Mediterranean chestnut plantations where, along slopes thousands of feet high, the land is held up by rock walls: here are the hanging gardens of Babylon, and the queen who built them is called the peasantry. The largest and most beautiful monument of Europe, where roofs and towers emerge from the trees and reflect on the waters, is called Tuscany, Scania, Touraine. What we take for eternal nature is the meadow, the edge of the woods, which go back to jungle if the peasant neglects them but for one spring; the riverbanks planted with great trees that a chainsaw can destroy in just a few hours. Their reflection will dance for all time in a Monet painting. This countryside is forever engraved in the eyes of one last generation, and we cannot imagine it disappearing, any more than invincible nature. And yet, more than natural ecosystems, it is the fruit of a fragile equilibrium, of an accord established at length between humans and their earthly surroundings. From the fjords of Norway to the islands of the Aegean, tourists believe they journey in nature when they are in fact walking through a garden that the toil of peasants has cleared in the primitive forest, leaving nature its share. But the industrial and 'urban' explosion destroys town and country along with nature, to mix them up in suburban chaos. This will soon be completed all along the coasts, the great valleys, the great roads and, above all, the highways of Europe.

If nature is habitable only when man carves out his share of it, for its part, a town is worthy of the name only when it leaves nature its share. Which is not to say that nature is confined to parks or squares, or worse, to that deceptive abstraction: green space. A town remains habitable when nature is present in it in an altogether stronger, subtler way. When the sky continues to open out above the town dwellers' heads and the purity of air and water and the nocturnal silence daily regenerate their lives. When the town, instead of dissolving into the haze of suburbs, remains rooted in its site and does not bury it under asphalt and concrete: its true 'green space' should be the surrounding countryside. Above all, for the town to remain a town or, better yet, a city, nature must invisibly pace its material and social life. Let a balance succeed the frenzied manipulation of developers, be they capitalists or city planners.

Man, this singular being, and nature ... Without deifying one any more than the other, man must give himself an environment where he feels no more lost in the forest than in virgin suburbs. Neither slave nor tyrant, he must make a contract with Mother Nature, after having signed one with himself: a planetary contract, which is yet local and personal. Frenchmen can still behold the beautiful remnants of a fine example of such an accord in the unique and multifarious landscape of their towns and the adjoining country. The countryside tells us what is the nature of the pact – which can be different and perfected – between freedom, progress, culture and nature. Its fruit is no more an asphalt jungle than a tangle of vines, for it is a garden – neither of the French nor of the even more artificial English style, to say nothing of a park managed by some ministry – in which the share of man is that of the earth. Eden already was one, they say.

Nature and freedom united in the existence of each human being

Both stranger and son to the earth: no reason can bridge this chasm, it would seem. For it is up to the subject to do it, up to me or to you, which cannot go without some physical and spiritual travail. But if, like the fairy tale hero, I walk towards the obstacle, it dissolves, and there is a lot of evidence that what is distinct can and must be united. But by a new bond.

It is precisely when nature becomes for me the Other (let us also say difference, to follow current fashion) that it exists for the existent that I am. When I no longer reduce my own being to nature by considering man as a mere element of the universe, or when I no longer reduce nature to man by considering it as the mere material of his works. When I no longer personify it like the ancients, or as Christians do by identifying it with a providence in charge of satisfying our needs or our reason. Or again, like some naturists, by endowing it with rational and moral qualities that are specifically human: such a naturism is also nothing but anthropomorphism. To know nature, you must distinguish it from yourself: love it for its own sake. To love matter as matter, for the silence and the weight with which it faces us, the plant for its rootedness and the beast for its innocent ferocity. You do not love it if you want to domesticate it, turning it into a kind of homunculus, like those women who own a dog because they cannot have a child. To be concerned about trees, or rather about this oak or that ash, there is no need to turn it into a tree spirit; the rustling of its leaves is eloquent enough. To recognize nature is to accept the whole of it, warts and all, down to this dark side of the earth – and so of man – that is given over to absurdity, necessity and death: practicing an *amor fati* that is the opposite of resignation to fate; whereas fleeing it by embellishing it amounts to a denial of it. The love of Nature with a capital N, identified with perfect Totality, is not love. To divinize it, and therefore humanize it, as pantheism or naturism does, is to reduce its tragic and fertile abundance to a bloodless concept: to an ideology, the exact opposite of nature.

Just as not accepting oneself as other than one's fellow is to hate and flee oneself, not accepting the other as other is tantamount to negating it. Love (and we cannot avoid this word made somewhat gooey by overuse) is the ability to discover some likeness to oneself in what is different. And this additional force, this passion that drives the spirit to go beyond its limits, is given to it by the body: you have to have a pretty weak nature to love only your own reflection, like Narcissus. For the *subject* that I am (excuse this title I give to all and sundry), nature is the *object*: the harsh reality that resists my material and spiritual grasp, the obstacle whose challenge sets my freedom into motion. A personal relationship with nature is agonistic, like any love relationship. If I love that nature whose existence exceeds any definition, it is because it gave me life, which is motion, and impels me to this other motion which consists in thinking and acting.

Beyond – not beneath – the contradiction that, once acknowledged, must be endured, freedom meets nature again. The burden of necessity is also the burden of existence, and its highest manifestation is but the gift of the earth to the species and of the body to the individual. Individualization is dependent on a physical and sensuous presence, active and reflective, which is never affirmed so much as within the silence and mutable expanse of the earth; whereas in town, the individual, submerged in noise, spectacle and social relations, is lost in a human environment he has trouble distinguishing from himself. The romantic stereotype of the solitary wanderer alone before the panorama is not false; it becomes false only if there is also a parking lot and a thousand tourists. The freedom of the city dweller is very likely to be negative and social; in the middle of the crowd, it mostly exists in his head. Whereas in the countryside, inscribed in space-time at all hours, it turns into a reality; which makes it all the rarer, because it is so difficult. But to the solitary wanderer – and still more to the local inhabitant, nature gives the joyful strength to stand erect in its midst. For the city dweller, going back to nature is touching base with his wellsprings: going back to the origin, which is not just material, to draw from the dragon itself the physical and spiritual energy that keeps an individual standing, ready to face society as well as the universe.

Moreover, if nature is the condition of freedom, freedom is that of nature. Nature is man, which can be understood in both senses, because if the universe constitutes man, it has to go through man's mediation. If there was not someone to know nature both carnally and intellectually, it would not exist. Let us not forget that each time we think and speak about it – and worse, when we turn it into an idea – it becomes a product of culture: the simple fact that we talk about nature shows that we are no longer part of it, and knowledge of it is the best way to limit the damage. To be logically consistent, an ecologist would have to remain mute like the plants and, as a matter of fact, some extremists go so far as to value autism. But in so doing, he would betray his mission, for only man can defend nature against man.

The nature that gives one the strength to be an individual can only be truly loved by him: we have seen how much that love is tied to the rise of individualism. Feeling – since it is customary to reduce to this maudlin term what has become a rage in our time – for nature, down to its harshest differences and its hostility to humans, can only be the attribute of a solitude: no sooner does it shift from the individual to the collective, as we see with tourism, it becomes the worst destructive force for nature. On the other hand, insofar as it remains personal, only this feeling can save nature.

It is not by renouncing his freedom and returning to the original jungle that man will preserve the earth, but by carrying this freedom to its own conclusion. At the point where his means give him the illusion of being able to break with nature, the time when he respected nature because he was powerless before it is over; from now on, he can only respect nature out of love, and nothing can force him to take such a decision. The old law has been replaced by a new one, at the price of anxieties and risks befitting man's true stature. A supernatural law; the antithesis of nature is not reabsorption into it as a result of a planetary disaster caused by the uncontrolled growth of the species, but the choice to master this growth. In a sense, there is no more antinatural act than that of rejecting the old law, cosmic and divine: 'Increase and multiply.'

Since 1945, the damage is done, as it were: man is cornered into being free. Considering the devastation already inflicted and the power at man's disposal, he has only one way left to avoid disaster: a supplement of consciousness which, after having dominated the world, should lead to self-domination. In other words, after nature, it is now the turn of the social supernature that exploits it to be subdued.

Nature and freedom associated in history

At first opposed, nature and freedom now end up as allies before a common adversary. From the origins, obsessed by the irresistible need to fashion the world after his own desire, man has given himself weapons, of which the most effective one has been association – which is not, as Rousseau believed, a free contract between individuals but a necessity inscribed in social being by nature itself. Alone, man remained weaker than the great predators; united with his fellows and accumulating the experience of previous generations thanks to language, he became invincible, at least on the scale of the planet. Early on, to counter natural disorder, he came up with the order of society, largely mythical at first. If he could not dominate nature practically through science, he humanized it magically through religion. He personified the disturbing inhumanity of natural forces under the guise of gods who, at first implacable, were still less so than the silence of space and infinite forces. Thanks to agriculture, he was able to produce surplus wealth that allowed him to build cities. Sheltered by their walls and their gods, in spite of famines and epidemics, in the middle of crowds and festivals, city dwellers could imagine escaping from nature and time. Then the State, that is, political and juridical organization, brought together cities and tribes, consolidating their power. Finally, Western science provided political and economic power with technical means that allowed it to exploit nature and control humans much more efficiently. For it was at the price of an even more thoroughgoing division of human activities, with their concentration and complexity growing apace; and today, in spite of dreams of self-management and regionalization, this process goes on like the very course of the stars.

Man frees himself from nature only by developing a social supernature – or second nature. He only escapes from necessity by submitting to other constraints, more deterministic ones, being less external than the initial ones: when man binds man, the knots are a lot tighter than if nature had tied them. And the more social organization progresses together with the exploitation of the earth, the more the freedoms it provides are paid for with new constraints. Accelerated development, and not just that of nuclear energy, is hardly compatible with democracy. Nothing of the kind can get done without rules and police; the freedoms obtained with the car are paid with many a discipline, especially in urban life. And each time we want to accelerate development, we have to suspend the legal guarantees that slow down its implementation and allow discretionary administrative and technical powers to prevail. Thus, to ensure the speedy development of a highway network and the land consolidation required by the mechanization of agriculture, the legal procedures put in place by the French Third Republic to protect private property and the interests of citizens have been discarded. The ZAC procedure (for *zones d'aménagement concerté*, 'areas of comprehensive planning' – oh so thorough!) facilitates expropriation for the benefit of developers, but it is for the construction of nuclear power plants that the greatest leeway is given to discretionary powers.

Since nothing on earth has ever been given for free, if man purports to achieve total emancipation from nature, it may well be at the price of total social control. That is why the threats of catastrophes resulting from a limitless exploitation of the earth are compounded by others, stemming from the reactions of human material crushed under the increasing weight of social machinery by the physical and spiritual malaise of man, who can tolerate happiness only as long as it is his own and not manufactured for him in research facilities or government offices. Such a paradise, lacking its share of nature and freedom, of body and spirit, can only be hell. But the very failure, both ecological and human, of this 'progress' can only result in its reinforcement. In order to control the dangers inherent in means that grow both in power and fragility, due to their increasing complexity, to manage diminishing space and resources, to anticipate and master human reactions that might get in the way of this, there is no choice but to intensify organization. It is therefore imperative to know, to calculate, to predict everything about nature and man. And since this phenomenon goes beyond all borders, only a world organization – a State, managing the whole planet by computer, would be in a position to avoid disaster. Unfortunately (or fortunately), it is far from certain that such an organization could be achieved, at least without wars and other calamities, due to the rise of nationalisms as they get exacerbated by the standardization of the planet. In any case, be it individual or collective breakdowns or natural catastrophe, chaos or a total planetary system, daubed in red or in pink, set up for avoiding it, it all comes down to the same thing for whoever cares at all for his land and his freedom.[1] At best, man would have extracted himself from a total nature only to surrender to another, totalitarian 'nature'. The first nature, having been the cradle of life, had given human freedom its chance, whereas the second nature would only be its grave. We may see the burgeoning of 'feeling for nature' in

---

[1] On this dialectic between system and chaos that keeps leading back to the system, see Bernard Charbonneau, *Le Système et le Chaos. Critique du développement exponentiel*. [Paris:] Anthropos, 1973 [Paris: Sang de la Terre, 2013].

the first industrial society and that of the 'ecological movement' in the second as a feedback, a warning signal that arises just where entropy threatens to overcome life.

* * *

Man cannot be reduced to a nature that, by giving him life, led him to the threshold of wisdom. Nor can he be reduced to this second nature that allowed him to emerge from the first only by burdening him with other chains and absurdities, compounded by a short-sighted reason. Only freedom, which is that of each one of us, will save life on earth from the blind will to power of the species, by profaning the second nature as it did the first. It is not from sociology that we should expect such a sacrilegious act; it is from you. But, as with nature, if man distinguishes himself from society – and so distinguishes it – it is to create other ties, born of a free contract, between its members, inspired this time by love and conscious choice, not by gregarious instinct. That day, at the same time as he preserves the earth, he will also answer the age-old human dream of an Eden where man is at peace with man as much as with tigers.

In so doing, it may be that man will be fulfilling an imperative inscribed in the very origin of nature: of evolution. That first disturbance in the inert serenity of matter: unconscious life, having won only a local victory over nothingness, it is now up to human consciousness to prolong its trajectory, running counter to entropy, by mastering the energy of society after that of the cosmos. This phenomenon, against nature and against society, that is individualization as the vector of freedom, is the step – to be sure still more narrow and fraught with risk – of a mutation as great as that of matter turning into life. It comes from just as far to try its luck on this tiny planet, and in this still more fragile human germ. The freedom of the individual may well issue from the innermost depth of all things. After all, it is not man who has created man. And it is nature that has invented sexual differentiation which, by multiplying individual differences to the extreme, multiplies the chances of another universe coming about.[2]

---

[2] See the series of articles in *Le Monde* (9, 10, 11 February 1979), in which F. Jacob shows the role of the differentiation of the sexes in the multiplication of genetically different individuals. For once, the ignorant may call on the highest scientific authority.

But every mutation is a leap, a change of reign. The step of freedom, in contrast to that of life, is free, and the one taking it can win everything or lose everything. If the price of life has been suffering and, worse still, the awareness of death, the price of freedom is an anxiety as great as its value. No wonder then that we are tempted to avoid this fearsome honour that leaves us alone to carry the earth and the universe on our shoulders. If man, sinking back into the social whole, were to one day destroy himself along with his home, this would only go to show that his freedom was but a myth too high above the reach of the anthropoid. And nature would have automatically rectified its error.

CHAPTER SIX

Nature and Christianity

The debate that opposes and unites human freedom to nature takes place at a metaphysical and religious level. It will remain without a conclusion in theory or practice so long as it is not carried to that level. And the ecological movement, ignorant of its sources, will be condemned to remain at the surface of facts and to act in a fog from one day to the next.

And yet, when it comes to its origins as well as those of the development it fights, Christian faith has played, and still plays, even unconsciously, a key role. The problem of nature and freedom is also found in that of nature and Christianity. And there too, the initial contradiction can only be overcome where it is endured and acknowledged, especially by the more or less rebellious Christians who have played a major role in the founding of the ecological movement.[1]

The rupture of Creation

Those rare ecologists for whom ecology raises the religious issue are reticent towards Christianity, and with good reason. They insist on the fact that the God of Jews and Christians grants man a

---

[1] On this topic, see Bernard Charbonneau, *Le Jardin de Babylone*. This thesis is more fully developed in Carl Amery's *La Fin de la Providence*, [Paris:] Éditions du Seuil, 1976 [*Ende der Vorsehung: Die gnadenlosen Folgen des Christentums*, Reinbek: Rowohlt, 1972].

privileged situation in the universe, which drives him to look down on and exploit the nature put at his disposal. According to Genesis, 'God said, "Let us make mankind in our image, in our likeness, so that they may rule over the fish in the sea and the birds in the sky, over the livestock and all the wild animals, and over all the creatures that move along the ground."' And after having created man and woman, 'God blessed them and said to them, "Be fruitful and increase in number; fill the earth and subdue it. Rule over the fish in the sea and the birds in the sky and over every living creature that moves on the ground." Then God said, "I give you every seed-bearing plant on the face of the whole earth and every tree that has fruit with seed in it. They will be yours for food"' (see Genesis I). Thus, man received as his property the earth that Providence created for him. But nowhere does it say that he has the right to destroy God's handiwork.

This sovereignty given to man has another, even more basic reason. If God gives it to him, it is because God created him in his image: sovereignty over nature belongs to the very being of the God of Jews and Christians. Greek or Oriental 'pagans' divinized it; the personal and transcendent God distinguishes himself from it. Being is only himself, beyond any material or intellectual, natural or social form: any formula, thing or institution. The One, personal God of the Old Testament is nonetheless pure freedom, as it were. Before the Creator's act, there was only an unfathomable night in which the spirit was dozing in its mother's womb, as the swell still rolls in the prologue of Haydn's *Creation*. Until the day when the lightning strike of a thought split the perfect atom of the primal Whole, in which God *distinguished* light from darkness, the sky from the waters, the waters from the earth. The divine cosmos was henceforth no more than a creation of the divine mind. The spirit became transcendent to matter; an invisible thunderbolt, volatilizing the sacred, suddenly petrified tree spirits and high places into trunks and boulders: into mere things. And man's knowledge and hands were able to lay hold of these objects that now seemed inert. To create matter, the Spirit had disincarnated itself from it. And this fundamental event is still going on today.

Yet at first, the rupture was not complete, for the Creator had placed his creature in a wonderful garden where it did not need to struggle against nature, and where, stark naked, it could give itself over to life with the innocence of children and beasts. But Adam and

Eve ate the forbidden fruit of the knowledge of good and evil and they were driven from Paradise, cast down into time and doomed to death, forced to give birth, think and work in pain. And to reach man, the divine curse extended to nature, which became a jungle ruled by the survival of the fittest.

Original sin consummated the rupture between Adam and nature, which he is compelled – and even has the duty – to struggle with in order to live. Which he doggedly does to the extent of his powers. Condemned to till the earth, he is less and less in magical communion with things, brought to mere utility by a will to power that reduces them to dust as soon as he lays hold of them. An ambivalent curse since it was imposed by God, work is both a duty and a blessing that happens to come along with the promise of deliverance from it. On the other hand, since there is no escape from it, we might as well make a virtue of necessity and fully enjoy the power nurtured by the pleasure of being at work on earth.[2] It is no coincidence that the Puritans, being the great readers of the Bible that they were, had a religion of work that they transmitted to capitalist societies.

Original sin had other consequences. It made man the enemy of a nature doomed to evil down to his own body, especially his sexuality that is its most intense manifestation. His existence became a war led against himself as much as against things. Any yielding to the delight of the instant, to ease, was suspect, and desire was severely reined in by will, religion, reason and morals. The New Covenant (of Yahweh with Abraham) only heightened this tendency. The religion of the God of the meek and the poor nailed on a cross made people sacrifice the enjoyment of riches and the pleasures of the flesh, which are primarily selfish, to good works performed out of charity for the neighbour. As for the curse–blessing of work, a switch happened. As long as we are going to bear suffering and inflict it upon ourselves, we might as well derive delight from it, either by enjoying other people's suffering out of sadism or our own suffering out of masochism: a specifically human and Christian vice, doubtless unknown in nature. But look at all those new pleasures!

What belongs to the body is not so much condemned as negated by the New Testament. Jesus, born of the Holy Spirit, does not

---

[2] 'Whatever your hand finds to do, do it.' Ecclesiastes, 9: 10.

recognize having a father or a mother, and his disciples are urged to cut all blood ties. Pleasure is limited to that of some modest agapes. If he provides the best of wines at the wedding in Cana and if he makes it the symbol of the Last Supper along with bread, he is neither for nor against Eros, he ignores it – unlike the Old Testament, much more involved as it is in flesh and matter. Jesus welcomes the adulterous woman and the great sinner but, to the best of my knowledge, he does not encourage them to continue their activities; and he recommends to his disciples that they become 'eunuchs' out of love for him. Except for Peter, whose wife is absent, this small group is a community of single men followed by a few women; and although he calls himself a bridegroom, Jesus had no other bride.

At first sight, it is the very spirit of Christian faith that drives people to break with nature, rather than such and such passage of the Bible. The nostalgia for Paradise lost makes people dream of another universe in which human desires would be satisfied without effort or work, where eternal life would follow upon death, and justice and peace follow war. Above all, the pursuit of absolute power and perfection fosters a permanent dissatisfaction that is soon disgusted with whatever it obtains. And anxious thought tirelessly seeks the means of rectifying not only the order of nature, but also the order of a society that reproduces it while betraying it. The call of a personal God to personal truth and freedom sows everywhere the seed of rebellion against the way things are, and everywhere makes it sprout; sooner or later, a prophet arises to denounce what had been reality, morals and the law until he came along. Unlike other religions, particularly Hinduism and Buddhism, that justify the individual's going with the flow of an invincible and sacred cosmos in which time eternally recurs, Christian faith, though an enemy of the sword, has nonetheless armed man with it.

From now on, time does not recur, history has been set in motion and humanity progresses: towards what goal? The further it goes forward, the less it knows. The moorings having been cut, evolution turned into progress can start, at first imperceptibly on a small cape of huge and heavy Eurasia. For, always for good or ill, the old man lives on in the new: the pagan in the Christian. One may not like it, but one must live on this earth, to get along with a nature that is still hardly mastered and the society that is indispensable to this end. And, purporting to freeze the time that Christian faith had caused to move, a thousand-year Christendom is erected which, revised and edited by Aristotle, makes the laws governing what is below match again those governing what is above. And the heresies and rebellions that come to disturb Order are duly repressed. To no avail, for the soil is fermenting. Thinkers and heretics criticize ever more boldly the Truth in the name of Truth, Justice in the name of Justice. Any principle undergoes a critique, any reality suspect of also being but an appearance is indicted before the tribunal of reason and compelled to produce its evidence. Religious, intellectual, scientific and political revolutions, crises and wars accelerate change which, from being local, becomes planetary. Transportation and communication machines churn and propel – more, always more – the maelstrom that our life is becoming. The human species used to go on foot and on horseback towards its unknown destiny; from now on it rushes there by rocket.

Incarnation

But it would be a mistake to reduce the Old and even the New Testament to a progressive ideology; Judaeo-Christian tradition is far richer, since its aspects, like its authors, are innumerable. There is hardly a chapter without its own retort: even the condemnation of the flesh finds one in the celebration of physical love by the Song of Songs. If Judaeo-Christianity can be reproached for something, it is not for being true or false, good or bad, but for being tremendously ambiguous: it is up to the reader to choose and seek in it what feeds him.

At the same time as the condemnation of nature, we find in the Bible its glorification. It is everywhere in the Old Testament, rooted (or mired), far more than the New, in its soil and its people: in the Promised Land that is not in Heaven but smack in the middle of a geographic and historical crossroads. The *Book of Job* celebrates Yahweh's omnipotence as manifested in his creation. *Ecclesiastes* enjoins man, because he is mortal, to enjoy the pleasures of the moment. And upon leaving the desert, it is not in a city but in a country paradise 'where milk and honey flow' that the Eternal leads his people. On the other hand, everything that strays from nature is suspect. If God has created the tree of knowledge, it is Satan who drove man and woman to eat its fruit; and to the end it will remain that of good *and* evil. The State, if not society as meticulously regulated by *Deuteronomy*, is what Yahweh and his prophets are wary of. If a king is needed for the people of Israel, even David knows weaknesses and Solomon goes bad – to say nothing of Saul and Ahab, possessed by evil spirits. Money, and worse still, the royal census, are works of the devil: what does Yahweh think of a national accountancy that ends up surveying everything with its computers so as to completely master nature and man? The Eternal grants his favours to Abel, the nomadic herdsman, at the expense of Cain, the sedentary agriculturalist and blacksmith, and it is the divine curse that condemns him to build the city. The Heavenly Jerusalem is not of this world, and things go awry every time man attempts to build it on earth. Babel, which wants to dominate the planet, is the work of an evil spirit, and it is annihilated for this reason. If the desert is the source of all virtues, Sodom and Babylon are the mothers of all vices. And the *Psalms* and the *Prophets* constantly renew the condemnation of any human work that wants to equal that of God.

'You are dust and will return to dust.' Ceaselessly, the Old Testament reminds man that he was drawn from the silt of the earth. The sovereignty he has been granted is not absolute like that of his creator, it is bounded by Adam's finitude, and due to sin, his work is never purely good. If, instead of being the vague sense of a general and abstract evil, the awareness of sin and evil was that of our own limits and of human weakness, it could be the wellspring of a more realistic view of nature, and warier of man and his works.

Although the New Testament continues the spiritualist and universalist tradition of the prophets, it remains nonetheless rooted in a Galilean countryside peopled by shepherds, agriculturalists and fishermen, where nature is omnipresent. The purest Gospel speech has the clarity of springs, the simple and sharp colours of spring meadows. Far from being cursed, nature is held up as an example to men, with their anxiety and greed for power and money. Christ asks them to draw inspiration from the birds of the fields that have no thought of tomorrow and know that in any case God will provide for their food, or from the lily of the fields which, without working, is still clothed more magnificently than Solomon in all his glory. But the glorification of nature in the New Testament is not exactly that of the Old. It is no longer its power that is praised, but its humble beauty and its carefreeness. What is put into question by the Gospel and the prophets, more than nature, is the social power that does it violence, as it does to men. It is war, money, the Law. Christ lives from one day to the next without family or trade, like an anarchist who ignores the economy and politics, without which men would have little power over nature. If Christians had strictly followed the Gospel's teaching, their power would hardly have gone beyond that of a tribe of gypsies or Indians.

Yielding to the spontaneous tendency of a supernatural and transcendental model, Christianity seems condemned to indefinitely bring upheaval to the earth in the attempt to realize an impossible ideal, as Gospel anarchism is condemned to subvert a society that can only realize the conditions of freedom by translating them into laws and sanctions. But if the old law is abolished, it is in favour of another one that belongs to personal conscience and love: which is as true for nature, as we have seen, as it is for the neighbour. If Christ finishes the process of disembodying the spirit, he re-embodies it on the other hand as no other religion has done, in a God-man who, through his body, lives, experiences death throes and then expires on a cross in his time and place. This reintegrates, albeit by a radically different path, physical and earthly existence, which a Christian would be tempted to look down on. 'He who hears these words and puts them into practice has built his house on rock.' The rock on which everything is built is not just the divine precept, but its paradoxical putting into practice; and no law determines how that is supposed to happen, it is up to freedom to do it. When that happens, nothing is negligible anymore: neither earth nor history; at every moment, a game is being played out in which the stakes are personal and universal salvation.

And the order of temptations which Jesus undergoes specifies their importance. The first one to assail man is that of bread: to magically exploit the earth so as to feed oneself – and, since this is inseparable, to feed one's fellow man. The second one follows from the first: it is the political temptation that is indispensable to the exercise of economic power. But the last trial is the specifically religious temptation: becoming God, or at the very least an angel whose wings save him from earthly gravity. Refusing Satan's offer, the son of God took the stairs instead of jumping from the temple's roof. Now that man has given himself material wings propelled by reactors and rockets, let us hope he will not forget the earth. Falling from such a height would be fatal.

The law of love follows upon the struggle for life and war, freedom takes over from necessity, but embodied in the existent par excellence: this man called Jesus, or John Smith. The creation – nature and destroyed society – is recreated. But if the old chains binding man to the earth and man to man held on their own, the new link can only be tied freely by every man, at the risk of losing himself. There is only one thing we can hold against pure Christianity: Is not the challenge it puts to the hominid mammal, that of a new Law embodied in an individual freedom, too far beyond his capacities? I have no idea. But this is no longer an academic question: the old order is crumbling and we do not have any other way. If it happens that man is not up to the challenge of his own destiny, then that will have been the mistake of his Creator, whether God or nature.

Christianity and the ecological movement

Progress, the continual development of science and technique, is inseparable from evangelical Christian faith; without it, it would have lacked an engine, nothing would have driven humans, until then steeped in the sacred, to break with the gods, except for a God-Man. The founding fathers of Western science are orthodox or heretical rather than atheists, and they are often tied to the most rigorous tendencies of the Christian faith: Franciscan mysticism, Jansenism or Protestantism. Roger Bacon, Leibniz, Pascal and Newton are Christians who are just more demanding than others but who have unwittingly unleashed an atheistic science they would certainly have disavowed. If the Christian religion eventually declined, its spirit and its values persisted, all the more virulent as no orthodoxy or church kept them under control.

A secularized, rogue Christianity is at work throughout the human species and the earth to establish within them, along with power, the perfection of happiness, justice and freedom. It no longer acts in the churches. 'It is more active in the Chamber of Industry and Commerce of the Federal Republic of Germany, in the Central Committee of the Soviet Communist Party, at the Pentagon or in the boardrooms of multinationals, than at the Vatican for instance, or at the World Council of Churches.'[3] Since the churches have lost their temporal authority, they are only thinking of one thing: adapting

---

[3] Carl Amery, *Fin de la Providence* [*Ende der Vorsehung*].

to their time, to technique, to liberalism or to socialism. This even though these are but by-products of their religion. Instead of yielding to the inferiority complex to which they are prone, they would be better advised to wake up to another feeling, even more unbearable though it may be: that of their responsibility for progress as unleashed by the ambiguity of Christianity. But such a feeling, before applying to an institution, starts out as that of lone men.

For this freedom of man that threatens to destroy him with his earth contains its antidote. It is in the very societies where the science and individualism born of Christianity have developed the furthest that feeling for nature, and eventually the ecological movement, were born. It is no coincidence that Rousseau is a son of the Calvinist Rome. The egalitarian, pacifist and libertarian side of the ecological movement, in which Christians who have more or less left the church are many, comes straight out of the Gospel. Its main weakness is that of a Christian idealism which can be reproached for fleeing (and therefore hating) reality; even if this idealism, to go along with fashion, claims to be a realism. And the consequences are always the same. Compulsion, hierarchy, violence are rejected. But the day we are faced with the necessity of acting, there is no more choice: if we do not want to be annihilated by the adversary, we pretty much have to organize like he does. And then the repentant idealist active in politics, convinced as he is of fighting the good fight, organizes, disciplines, hierarchizes and wages war even more efficiently than the enemy, thus emptying his action of content, the means causing him to forget the ends. And victory only ends up reinforcing power over men and nature; this has been the fate of all revolutions until now. In any case, Christian or post-Christian idealism only leads to failure. At least Jesus laid his cards on the table when he said: 'My kingdom is not of this world.' It is up to those who live in it to make do with this sole piece of advice.

The ecological movement can only know the world it refuses and know itself if it faces the Christian fact. But it will not be by reproducing it negatively in a systematic anticlericalism or anti-Christian stance: such a primitive reflex would only mean that it has been unable to gain any distance towards it. For a free spirit, it is just as pointless to reduce the negation of nature to Christianity alone as it is to deny the latter's role. It is more the result of the combination of the old man and the new. The former has not changed since Neanderthal, he still worships material force, but whatever little freedom Judaeo-Christianity has awakened in him has allowed him to dominate the earth. The inability of present man to discover the absurdity of infinite development in a finite world, as much as a product of the Christian passion for the absolute, is that of a past when human power was limited in a world that then seemed unlimited; and it is Magellan, a good Portuguese Christian, who first discovered the earth's limits.

The naïve anthropomorphism that causes us to view the planet as a table inexhaustibly set at mankind's disposal is a fruit of the lack of consciousness more than of conscience, which I doubt was any sharper in pre-Christian societies. The awareness of a world from which providence is excluded is hardly a trait of peasants and pagans. Concorde or the H bomb is the work of a Pithecanthropus who went to Polytechnique or Harvard. But Christian idealism has not dealt enough with the explosive mix it forms with real nature, real history and real man. Among other calculations of limits, it should have thought on its own instead of revelling in theological commonplaces about sin, be it by embodying them in some left-wing or right-wing political devil. Was not the disaster embryonic within Christian nihilism and providentialism already present in Homo sapiens? Had not the rot set in with his very advent? The enthusiasm with which all the non-Christian cultures of the planet have embraced the West's political ideology and technology may lead us to think so. The fact remains that without Western Christianity, this phenomenon would never have been so sudden or so extensive.

Above all, Christians make the mistake of denying, apart from their religion's ambiguity, the risks it has brought along with the step of freedom and love on earth, as its price. This being said, on the condition of going all the way, that is beyond, Christian faith answers the questions it raises. More than the creation, the incarnation paradoxically gives back to matter and the body their full spiritual importance. And faith alone will be able to impose the asceticism that takes their limits into account in all areas. We may say with Carl Amery that, since the sacrifices needed to save the earth and man 'can hardly find justifications in our immediate interests, the call to a religious renewal seems well-founded'.

> All things considered, the contraction of the earth brings mankind's situation back to that of Job. The lived experience of human and earthly finitude teaches man, and with what

thunderous noise, that nothing belongs to him, absolutely nothing, not even his last shirt, and that there is nothing he owns that he does not have to share – not just with his fellow sufferers, but with everything that lives, that is, that changes and disappears. This may well be, unless I am mistaken, a theologoumenon. And, in my opinion, there is only one way for man to honour and respect it. We have to profane, to liberate in its turn this last field of action and thought: we have to treat the future itself 'as though' it could and should be defined in purely human ways ... And we must not allow any agency, be it divine or human, to leave half-open the least way out, to count on any miraculous intervention whatsoever, to spare us the sufferings we have laid in store and inflicted upon ourselves with our own hands. We must, to speak in theological language, tend towards this final kenosis, this ultimate self-emptying: the renunciation of any guaranteed future. It is only by losing it that we will win it ... We have entered a new phase of divine unfathomability.[4]

In all areas, the time for a mere 'Increase and multiply', which was valid for the weak and rare tribes of an empty earth, is over. What is demanded of man may be called in Christian terms a conversion that can only be personal. He need no longer look any further than himself to a cancelled Providence, nor even to reason alone to help him effect this unprecedented reversal. 'We are summoned to understand the election of man as responsibility – without which he is nothing.'[5] He is free, like Atlas as he carries the earth on his shoulders from now on and, above all, free to refuse this crushing load. From now on, we find ourselves grappling with this Creation that its creator has entrusted to us. A Christian can answer such a challenge only by effecting a Copernican reversal at the level of religion itself; if it puts Christian faith into question, it does seem true to its general direction though. The current crisis finds us fundamentally involved in the earth which we had purported to escape. And it is no longer from the heavens or from nature or from history that rescue will come, but from the – paradoxically spiritual – experience of an earth where man for ever more makes

---

[4] Ibid.
[5] Ibid.

a decision against entropy, death and necessity in a struggle that may be crazy but is the only meaningful one. Only the freedom that is its conscience will be able to save us: this time in the sense in which we say that we save ourselves from drowning. But it is written somewhere that the spirit became incarnate in a body.

Since we are talking of the Christian religion, perhaps we should end by invoking the almost mythical work of a writer who has glorified nature by drawing his inspiration straight from the Bible. I am referring to Moby Dick, the whale with which Melville symbolized nature before the Friends of the Earth, which is pursued by the implacable Captain Ahab – whose very name is ominous. The white whale is depicted as a monster, a Leviathan, invincible and murderous. But is it really? Nature is not haunted by the rage for omnipotence; it is Captain Ahab who is the furious monster.

Now that he commands a fleet of factory ships armed with harpoon guns, the game is becoming unequal and Moby Dick is a poor excuse for a monster, facing extinction as he does. But it is just an illusion, for the more Ahab's fury grows, so does the monster he is chasing. And now, perched without knowing it on the back of Planet Life, he risks his skin by thrusting ever deeper in it countless darts, some of which are atomic. *Eritis sicut Dei*[6] ... a redoubtable promise inspired by the Devil as much as by God. Might this not be the true monstrosity, mother of all monsters? As much as to the Greeks' measure that gives back to every thing its proper dimension, it is up to the God-man to remind man of what he is. Unfortunately, if the divine God who rises again (among other exploits) comes naturally to us, the God who is born, experiences death throes and then expires like us is a paradox, humanly unthinkable. The Incarnation is a mystery. And until now, Christians have succeeded as little as pagans in unravelling it.

---

[6]'You will be like gods.' Translator's note: The quote is from the Latin version of Genesis 3:5, where the word is '*dii*' rather than the incorrect '*Dei*'.

PART THREE

Diseases and poisons: Contradictions and shortcomings of the ecological nebula

The fruit is ripening, the seed is germinating, woman and the earth are pregnant. In silence and at their own pace. This is the way of all creation before the harvest; it evolves only if it involves itself in self-reflection. And when it blossoms in broad daylight, all the winds of the sky assail it. Slowly driven up from the depths, the mountain arises, and meteors sculpt it, giving it its profile. The meaning of any work thus emerges from the erosion of critique.

CHAPTER SEVEN

Nature, freedom and the ecological movement

Any truly creative movement is a fruit of the spirit (it matters little whether the spirit is divine or human). Yet the 'ecological movement' will lack this spiritual foundation as long as it will not have overcome – and thus have posited – its basic contradiction between nature and freedom. Since it has not yet frozen into a monolithic ideology and organization, until now the promoters of tribe and family coexist in it with members of the MLF and out-and-out anarchists. But such an agreement runs the risk of not holding up when faced with the demands of reflection or the choices of action, for this theoretical debate has far-ranging practical consequences. If the ecological movement, caught up as it is in the immediate and in politics, fails to initiate it, it will remain confined in superficial confusion and one day, as circumstances demand, it will split between a right-wing naturist fundamentalism and a left-wing libertarian fundamentalism. Whereas if it holds the two ends of nature and freedom, it will be able to go forward on its own path.

The temptation of naturist fundamentalism

In spite of Rousseau, nature is on the right; already Burke and de Maistre reproached revolution for ignoring natural, divine and human laws. Freedom is but a sham if it fails to take into account the necessities ruling any reality. The universe has nothing to do with the narrow-minded desires and crazy dreams of the humans who are forced and duty-bound to conform to its laws and accept its mysteries. What we take for its absurdity and its imperfection obeys reasons deeper than our own: suffering, death and war are the price to be paid for life. The part only exists in relation to the whole: man in terms of cosmic order, the individual in terms of society. And just as necessity comes before freedom, what is has priority over what could be – and even more over what should be. The existent is worth more than the idea, what is established by tradition more than the u-topia that dreams of the future.

The no-doubt temporary triumph of culture over nature has resulted in the retort of a naturalism that opposes nature to culture in all things. But too often, just as the Right retorts to the Left and vice versa, this naturist fundamentalism reproduces to the least detail the progressive matrix that spawned it. Repressed by the religion of the Left dominant until now in the intelligentsia, it is reduced to seeking refuge in a quasi-clandestine ideology or the ghetto of a romantic or post-romantic literature. Naturist fundamentalism belongs to little-known but sometimes influential theorists such as R. Hainard,[1] biologists and naturalists whose specialty inclines them to emphasize nature and life. But since the failure of Nazism, it is difficult for them to openly express judgment on human societies that they draw from their knowledge of animal societies. The condemnation of Progress and Western culture is most often issued by novelists such as D. H. Lawrence, who have the artist's privilege of being left-wing while being right-wing. The anti-Christian naturist can also refer to Nietzsche, but let him beware, for with him the against is never far from the for. This fundamentalism, until now driven to the margins by the reign of the progressive intellectual Left, nonetheless plays an important role in the ecological movement's groupuscules and sects. Even within its Left, there is nothing to stop one from being an anarchist and believing in orgone energy and parapsychology.

The triumph of Progress is that of positive, mechanical and quantifiable reason. Its opposite is therefore irrationalism: mysticism

---

[1] Translator's note: Robert Hainard (1906–1999) was a Swiss artist, naturalist and writer.

and magic. A whole modern anti-modern current aligns itself with them as it invokes a sociology of the sacred and the Freudian unconscious, which give it scientific credentials. And many poets and peddlers supply the public's nostalgia with more or less efficient substitutes.

Returning to nature means finding again the sacred ties that unite man to the cosmos by doing an about-turn on the way that has led from Christianity to rationalism. After D. H. Lawrence and so many other intellectuals, some ecologists are haunted by nostalgia for a religion that would reintegrate man within the Whole as it resolves the contradictions feeding modern anxiety. But this pantheistic paganism, brought back to the boil by the Gospel's fire, has nothing to do with Greek measure and harmony, but only with Dionysus after his return from Asia. For this irrationalism, reason only leads to senseless material practices or to soul-destroying, sterile criticism; it is not consciousness, but the unconscious that opens the way to Knowledge. Hence the penchant of many young ecologists for magical and exhaustive solutions, somewhat disguised as science. Hence also the – often disappointed – search for the guru who will provide the universal panacea. Or, for lack of anything better, the resort to sacred poisons as a source of divine rapture.

Just as personal God and personal man dissolve in the cosmos, the individual disappears in the sacred society whose model is provided by historians or, better yet, by ethnologists. Communion, and hence the recovered rather than invented ritual that ensures it, takes precedence over withdrawal and individual reflection.

The more so as we ignore time, we dream of ceremonies and sacrifices involving the flesh and blood-letting to various degrees, which are supposed to save us from the anxiety it involves. We seek to escape it in the eternal instant of a festival we would like to experience on a daily basis. For this naturist fundamentalism, just as man is only an avatar of the cosmos, the individual is just an appearance – the true being is society, from which such naturism expects the bread of the spirit as much as that of the body. His sufferings and his death are thus of little import compared to true life, that of the group mediating between him and the universe. He should therefore accept the 'Increase and multiply' which, combined with death and war, ensures the continuance of evolution by selecting the strong at the expense of the sick and the weak. He too must sacrifice his life or take someone else's if the group's future demands it; and obey every day the authorities and powers that define the law and determine everybody's place and function, as in ecosystems.

The celebration of nature without the counterweight of freedom leads sooner or later to a biological determinism whose outcome is an 'ecofascism', which incidentally would have more to do with Nazism than with Mussolini's tragicomedy. If we put nature above human reasons and values, we are brought back to *Blut und Boden*[2] and to the struggle for vital space, that is territory. Confusing ethics and ethology, the nature fundamentalist goes so far as to cite animal society as a model for humans. Either that or the microsocieties still subsisting in an invincible, nearly intact nature: Prairie Indians or those of the Amazon forest, Eskimos of the Arctic. In spite of the favourable accounts of ethnologists, we may wonder if a Westerner could adapt to them if it involved more than just passing through them. The status of women in these societies is not exactly in accordance with the MLF ideal; nor are the bloody practices imposed by religion or by scarcity in accordance with the ideal of non-violent ecologists. As for the non-conformist who would purport to touch the customs and refuse the authority of shamans and the power of chiefs, he would probably pay for it by being excluded or even pay with his life.

Although some fundamentalists of nature view the peasant as its first destroyer, which happens to be true inasmuch as he becomes a mechanical or chemical industry worker, others are fond of taking traditional rural societies as a model. But it is sufficient to be old enough and to have known them otherwise than through bucolic accounts to understand that the price of the virtues of the Breton or Basque Catholic tribe – led by its notables and its parish priest – was heavy conformity, with hypocrisy to boot. There too, the status of women, most often reduced to a clandestine form of power, and that of young couples, was not always rosy. This type of society demands that we accept or reject it as a whole – this has both advantages and disadvantages. The individual who melds with it experiences a kind of collective dream that reassures him and makes him bear its material indignities, but a critical, independent mind has no place in it. He may at the most hope

---

[2]Translator's note: German for 'Blood and soil', a Nazi slogan.

to be tolerated in the capacity of a fool, or of a tourist when the village starts to open up to progress – that is to say, when it starts decomposing.

Today, ecofascism is hardly being heard from; it takes on a left-wing garb and comes down to the denunciation by leftists of everything they do not like. But it is too much a part of the logic of naturism to be no more than a label that can be pasted onto anything. The idea of a nature perfectly distinct from culture suits a society that practices the division of functions and zoning. Ecologism is the most fitting ideology for the handful of scientists and civil servants in charge of managing the tiny sector of a chemically pure nature from which man – but for the licensed naturalist – is excluded.

The nature fundamentalist can be seamlessly integrated into the industrial system as a manager of nature reserves or of national parks (the regional Luna Park hardly being more than a bastard of green space and trade fair) that serve as an alibi for industrial, real estate, land or tourism reserves, at a ratio of one lark for one horse. Within those few administratively frozen relic spaces, the naturalist can assuage his passion for pristine nature as the ethnologist can assuage his own for tribal societies in other museum reserves. But between a temporarily reserved nature (while waiting for the creation of the next ski station or military camp – more popular with naturalists because it is forbidden to the public) and the culture of concrete in asphalt, what the Frenchman is going to be deprived of is the countryside where the agriculturalist inhabits and preserves the earth for everyone. Being neither beast nor angel, neither bear nor ecologist in charge of studying it and housed on national park grounds in this capacity, I can only refuse a society that forbids me to inhabit my fatherland: the earth.

Despite appearances, ecofascism has the future on its side, and it could issue just as well from a left-wing totalitarian regime as from a right-wing one under the pressure of necessity. For governments will increasingly find themselves compelled to act in order to manage dwindling resources and space. Exhaustive accounting will record, along with all the costs, the formerly free goods used by both industry proper and the tourism industry. The sea, the landscape and silence will become regulated and manufactured products, to be paid for as such. And the distribution of these essential goods will be regulated according to the case by the law of the market or by rationing tempered by an inevitable black market. It will be possible to ensure the preservation of the oxygen level needed for life only if another vital fluid is sacrificed, namely freedom. But as in wartime, the defence of the common good, of the earth, will be worth the sacrifice. Already, the action of ecologists has started to weave this net of regulations, matched with fines and prison terms, which will protect nature against its uncontrolled exploitation. What else is to be done? What awaits us, as during the last total war, is probably a mixture of technocratic organization and return to the Stone Age: the intuitions of science fiction are likely to be closer to the coming reality than Mr Fourastié's progressive forward planning. We are going to have the Merovingian period managed by electronics: scarcity, violence and terrors that can only be made bearable by the power and sacred authority that will save the planet – or finish it off. Until the end, down to the smallest detail. It is up to this generation to prevent the resurgence of the year one thousand in the year two thousand.

A critique of ecologism

There is nothing like a half-truth taken to the limit to become the worst of errors. As much as ecology can be a gain for thought when it reminds man that he is not everything, and when it reminds a production-obsessed society of its impact on the environment, when it becomes an ecologism, it turns it into an ideology just as abstract as that of Growth. Because it too forgets, not a value, but an essential fact: man.

For it is there, starting in the naturist who makes nature speak: he can no more refuse man than get out of his own skin. The nature fundamentalist who purports to judge him from an external point of view by accusing him of anthropocentrism falls into this quirk in turn, as does a theologian who purports to judge men from God's point of view, though here it would be more like the Whale's point of view. We can talk of nature and man only by using the latter's reason and words. If the ecologist talks of nature with such warmth, it is because he loves it, which generally induces him to endow it with our reasons and our virtues; if he gives in to this tendency, he reduces it to a kind of manikin fashioned after our own moral ideal. It is inevitable that man would judge from man's point of view; we might as well know it so as not to be fooled by our own temper and social milieu. This bias in favour of one's species is only natural; any living being on earth does the same and carries its powers as far as it can, but without saying it or raising the issue of their limits. Besides, if we are going to take a distance from our own species, why not take it from its planet and from life itself? This viewpoint is just as subjective; if the ecologist could understand the non-language of white-hot or ice-cold matter that makes up the better part of the universe, he would hear himself be called a narrow-minded villager and a racist earthling. Man and his difference ... when you are the former, how could you fail to care for the latter? Still, this is no reason to defend it with atomic claws.

Man is there, with his specific nature that is a social or personal supernature, for his misfortune or his salvation. And after all, if he is a mistake, since he is not self-creative, it is nature that committed it. The key to the problem is not in nature or in man, but in their relationship, especially in a space as profoundly humanized as that of the Europe of cities and countrysides. Ecologism knows only one means for resolving the contradiction between nature and man: eliminating humans. Robert Hainard is logical – if not reasonable – when he suggests that we should give the countryside over to the wilderness, where only naturalists would make a foray once in a while, while the bulk of the population would be withdrawn into what sounds like urban space capsules, where it would live and be fed by artificial means. One can imagine the neuroses this break with the earth, effected in its name, would foster; besides, today already they prosper in our urban monads. When it comes down to it, the integral naturist has only one solution to offer man: a nature reserve extended to the entire planet. And at the end, leaving for Saturn or the suicide of the last disturbing element – the director of the Muséum national d'histoire naturelle.

Since the safeguard of nature can only be organized by man, ecological science deals him only some of the cards. The ecologist runs the risk of losing sight of the realities of culture, psychology, economics and society – hence of politics. And nature will always be the first to suffer, even if he rounds out his naturist stereotypes with leftist stereotypes. Nature's first good fortune is man's knowledge of himself, which also means that of the second nature, now replacing the first one almost everywhere. It is therefore not ecology but a sociology that is able to take a distance from the social fact. And since it is only an individual who can take this distance, science is not enough. The active awareness of nature is a matter of morals, or rather of ethics: of an awakening of the spirit. The protection of nature is rooted less in matter than in freedom.

The libertarian temptation

The tone of the ecological movement is mostly given by a libertarian left which can for its part be faulted for sidestepping the problems freedom raises for nature and society. Reacting against the industrial bourgeoisie's cult of (quantitative and material) 'facts', it values u-topia and also u-chronia as means of escape from earthly realities. While protesting the ransacking of nature, resources, landscapes and cultures as they exist locally, it calls for the removal of any constraint applied by society on its members, even minors, delinquents or mad people. It tends to confuse relative repression as it is practiced in Western societies with the absolute repression of some totalitarian regimes or the total – if not totalitarian – internalized one that electronic and genetic progress could well have in store for us. In this, ecological leftism falls prey to a fateful misunderstanding; it demands total freedom even as the industrial system in both West and East threatens in countless ways to totally negate it: it demands a Brave New World when what we need is to avoid the worst of all possible worlds.[3]

The left-wing ecologist's utopia is in fact a very old one: it is the old dream of a freedom that would be given in its entirety to everyone. To use a language that is beginning to go out of style, let us call it self-management, regardless of class, race, sex or even age, mental health or morality. The libertarian spirit is egalitarian: whatever their differences, wrongly called vices or virtues, all humans are entitled to the same freedom, which is one and indivisible. Equality in difference is an excellent principle; the difficulty starts when it is put in practice. Saying that women or even children are equal to male adults and giving them the same rights comes down to saying they are the same by attributing to them the same virtues. Like any

---

[3]Translator's note: This is a word play on the French title of Huxley's book: *Le Meilleur des Mondes* (lit.: 'The Best of All Worlds'), translated by Jules Castier (Paris: Plon, 1932).

left, leftism can be faulted for negating natural or cultural realities in the name of theoretical equality.

It is remarkable that the ecological left refuses to acknowledge differences as natural as those of age or sex. Which no doubt explains the casualness with which it treats *birth control*[4] and the issues raised by the abolition of the old 'Increase and multiply'. Like other freedoms, we may deem it necessary but it still goes against the grain of nature; it is hard to call abortion and the pill natural products or a natural act. As for the difference that opposes and unites man and woman, there is scarcely any other that is so much part of the very fabric of life, be it only in their capacity of genitor and genitrix or, dare I say, father and mother; and it is scientific artifice alone that will be able to abolish it. By negating difference – I am not saying inequality since we are talking about incomparable qualities – are we obeying a passion for freedom and equality or, on the contrary, are we being fooled by the development of a society that is becoming a vast factory for the production of standardized substitutes? If it erases the difference of yin and yang (to use a language that will appeal to the zealots of Eastern wisdom), how can all other differences be expected to fare? To say nothing of the salt of life.

The same goes for the liberation from sexual taboos in the name of desire or of 'libidinal revolution'. We are here at the heart of nature and the mystery of life (no need of a capital L), which would deserve to be approached with more caution, if not reverence. The practice of anything-goes-any-which-way in the name of desire, the putting on the same footing of pederasty or paedophilia with the love of woman and man is a good illustration of the tendency to push to the limit the logic of freedom. Which is hardly the way of nature, or of most traditional societies. There too, in the belief of being a threat to society, are we not just helping it along the road on which it is already hurtling down? In this area, as shown by the example of American Puritanism, it is easy to switch from hypocritical silence and the repression of sex to advertising that adds the spice of Eros to every commercial or even political slop.

Man and woman, father and mother, lover and mistress, parents and children. It seems hard to deny that these are primarily natural

---

[4]Translator's note: In English in the original.

differences which are merely amplified by the cultural environment. One is born a man or a woman, and then, according to the necessities or the accidental circumstances of life, one becomes it; society's share does not cancel that of nature. The negation of institutional or de facto marriage, denouncing an egotism for two[5] in the name of a broader erotic relationship, the call for children's freedom from the authority of parents lead to a condemnation of the family. Among ecologists, it usually takes place without debate; those who do not agree are content to remain silent. This negation that goes without saying in certain sects is likely to be as short-sighted as the old religion of family. If it is to be replaced, this would at least require a reflection that can rely on examples. And even then, one may wonder if the questioning of the familial hornet's nest[6] is being confused with that of its decay under the pressure of a political–industrial system that tends to bring every social reality down to itself.

A critique of the ecological movement's anarchistic and non-violent strand

This current's weakness remains that of Rousseauism more than of Rousseau, who, despite the goodness of human nature, was in practice rather realistic and pessimistic in his assessment of humans. The ecological left gets rid of the contradictions between nature and society on the one hand and freedom on the other by assuming nature and mankind to be endowed with all virtues, so that it would be enough to let individuals freely associate with their peers to re-establish, along with social harmony, the harmony of man and the cosmos. A theoretical optimism that turns into pessimism under the impact of experience. As in 1793, since these people who do not[7] correspond to the ideal projected onto them are not human beings,

---

[5]Translator's note: In the original, '*égoïsme à deux*' is an allusion to the *poète maudit* Lautréamont's cynical definition of romantic love.
[6]Translator's note: In the original, '*le noeud de vipères*' (lit.: 'the nest of vipers') is an allusion to a novel of that title (Paris: Grasset, 1932) by Catholic writer François Mauriac (Nobel Prize 1952), who looked to his own Bordeaux milieu to illustrate the meanness and hypocrisy of the bourgeois family obsessed with patrimony.
[7]Translator's note: The negation is missing in the original, but this is likely a mistake.

they are monsters we must coerce and, if need be, annihilate by any means available. Let us recall that, be it vis-à-vis either society or nature, man is neither absolutely a serf *nor* is he absolutely free. Freedom and order are only contradictory if one or the other claims to be absolute. The surest way not to introduce relative freedom in social reality is to refuse in the name of self-management any sacrifice to common work, law and institutions, which in many cases are reinvented more than they are imagined. The rare communities that succeed are those that forego permanent festival under the guidance of the Holy Spirit and instead give themselves a framework and laws, which are instituted by excluding, by force if need be, all those who do not accept them. And generally speaking, the coercive institutions that allow the economic and cultural community to exist under various names have to do with the couple, the family and the village. If they are not annihilated by an outside intervention, they are the ones that will regenerate the countryside. For it is almost as difficult to manufacture society as it is to manufacture nature.

The same goes for the State: to free ourselves from it, we must first acknowledge it. On this point, the ecological left is divided between the supporters of an 'everything is political' borrowed from left-wing parties and the opponents of parties and the State. Christian or post-Christian ecologists take up the Marxist explanation that brings back all the sorrows of the earth to this *diabolus ex machina*: Capital identified with Property. Hence the Solution: socialization. Unfortunately, in the East socialism has only accelerated the ruin of nature at the expense of freedom, while in the West, the technocratic State, with expropriation at its disposal, is the ally and promoter of Capital's designs. The 'socialization of nature' advocated by ecologist Philippe Saint-Marc,[8] former president of the Mission interministérielle pour l'aménagement de la côte aquitaine (MIACA), is a contradiction in terms. Society is the antithesis of nature, and the latter has nothing to do with its management. The case of the development of the Aquitaine coast shows that it is even more efficiently destroyed by the coalition of senior civil servants and urbanists than by the promoters they attract in order to realize grandiose plans, inspired by the need to

---

[8] Translator's note: Philippe Saint-Marc, *Socialisation de la nature* (Paris: Stock, 1971).

become famous as much as by greed for profits. When it comes to the destruction of nature – nuclear plants, military camps, highways, land consolidation and so on – even in the West, the State is always front and centre. As for cultural differences, they are abolished by public administration, the school system and barracks as much as by corporations.

Thus, the anarchistic strand is not wrong to mistrust the State. But I fear once more that it confuses the struggle for absolute freedom with the one against the absolute State: in the first case, one fights to obtain everything, while in the second case, one does it in order not to lose everything. And if the struggle against the State has been going on forever, its suppression is but a dream that until now has only led to its reinforcement in the name of a withering away that we are still waiting for. Anarchy has done little more than throw bombs and rule literature and songs. The governance of any society of a certain size, especially if it is equipped with technical means, takes place through a State, which is centralized or, at the very least, federal. One might as well know it if one wants to 'confine'[9] the central power within proper limits. The issue is not replacing the State by generalized self-management, but preventing the advent of the total administration. This can only be done by awakening at the grassroots level people and societies that will resist its grip, and by defining the faith and the common institutions that can federate individuals and societies that happen to be different. Certain truths and necessities proclaimed by the ecological movement, be they national, continental or planetary, are precisely of that nature.

Forgetting nature's lessons, the ecological left also dreams of a national and international society identified with Freedom and Justice, where any power relations and struggle for life would have disappeared. In the meantime, every moment it keeps on crashing against this wall; furious over the failure of non-violence, it becomes tempted by terrorism. There comes a day, in Malville or elsewhere, where one is forced to choose between being annihilated and fighting. Either that or it is only the enemy's violence that is called violence, and with some embarrassment, one ends up, in words at least, siding with the Baader-Meinhof Gang or the Red

---

[9]Translator's note: The French verb '*cantonner*' also suggests a play on the word '*canton*' for a small local community.

Brigades. Which is the height of contradiction for an advocate of non-violence.

In spite of the human desire for peace, as in nature, but in a far less regulated way, violence is everywhere in society. Social relations are in part power relations. Were it only for class struggles! There are those, even harsher, between churches, nations and trades, clans and families, and even individuals. Similarity and comparability induce rivalry; difference leads to war. Aggression is inherent in life, even weeds are imperialistic; and it is not a miracle drug that will rid us of this virus, but its recognition around us and, above all, within us. Provided we do not give this factual acknowledgment the authority of a value judgment, thus repeating in reverse the mistake of the idealist who takes his value judgment for a statement of fact. The problem is not that of choosing between non-violence and violence, but of knowing which is which and of mastering it to the extent that it can and must be mastered. Failing this, one gets mired in contradictions and dead-end situations.

Relations between States whose raison d'être is peace and war are even more than others power relations, and there too, it is not by negating war that we will end it. If we are to believe non-violent ecologists, conflicts between peoples that are naturally respectful of their differences are only due to the dark conspiracies of a handful of capitalists who prevent them from going along and establishing the world federation where the common law will respect everybody's freedom. And yet, pacifism has largely shown what it was worth during the First World War and, *bis repetita placent*,[10] during the Second World War, in which pacifists were to be found in the ranks of both armies in the field. Such lessons should at least have called for reflection. All the more so as nowadays it is no longer the ideal but reality that corners us into refusing, not so much War in itself, but this war, which is absolute. Indeed, this time around, it is the ultimate weapon that makes war into an absurdity as a matter of fact. For either it reduces war to small ones where atomic powers face each other through satellite States to avoid their own mutually assured destruction, or it draws them into an atomic conflict that may well annihilate the human species. In this case, any society that contemplates it by giving itself a means that implies the End

---

[10]Translator's note: Latin for 'That which pleases is twice repeated.' A quote from Horace, *Ars poetica*, verse 365.

becomes the Enemy of every man. It always comes down to the true raison d'être of the ecological movement: not establishing paradise on earth, but preventing hell from taking hold of it.

Beyond the ecological right and left

Faced with the destruction of nature that, within such a system, only that of freedom can prevent, the ecological movement must refuse the polemical debate pitting against each other a conservative right and a libertarian left even within its fold. On the contrary, it must come to an agreement on a defence of the earth that would be that of man. By ceasing to oppose one to the other while remaining faithful to itself, it will prove realistic for a number of reasons. For, by uniting nature to human freedom, the ecological movement gives itself at once the two principles that make it possible to explain all the misadventures of the present world and the two most powerful motivations that can move the human spirit. It is no longer a half-truth that it is putting forward, but total life and total faith. But by confining itself to one or the other, by depriving itself of half of its reasons, it risks losing most of its potential supporters: for instance, if it defends plants and animals by refusing to open itself to real agriculturalists, hunters or fishers, who are its natural allies.

In this era of the death of God, nature and freedom – belonging to the earth and the right for every man to be himself – can alone found a consensus that is at once personal and universal. Any other one can only give rise to war between individuals, classes, nations or churches. By contrast, the fact that our body perishes if air and water are poisoned, and the fact that our spirit dissolves if the individual is denied any identity and existence of his own, these are the two ultimate truths. And these two are really but one, which, on these two feet, stands firm and becomes able to move.

Man and his earth, distinct and associated in the former's consciousness and work, constitute the sole foundation of a political action and a social project that will only be able to save one along with the other. And it is the tension between these contradictory terms that are yet indissolubly united, like body and spirit, that makes up the wealth and life of the person or the community that joins them together. For it is no longer a ready-made truth, the logical automatism of a monistic ideology that generates the certainties needed for action, but a constant critical return to these two poles that are as those of the same planet.

Man is nature and freedom as he is society and individual – I believe that at the endpoint of a necessarily insistent demonstration, one may refrain from underlining the conjunction. When man denies this in one way or another, he decomposes. Without his freedom, nature is not even a word, and his freedom is but a ghost when it refuses its carnal and terrestrial body. But by going all the way to its own end, it accepts it; then spirit becomes flesh and a second Genesis begins.

CHAPTER EIGHT

A fruit still green

Some activists of ecology will no doubt say that the fundamental issues thus laid out are devoid of interest: since everything has been said, all there is left is to act, that is to 'do politics'. In this, they take up the central idea of the society they denounce, namely that thought is the opposite of action.

Ecology without a doctrine

We know all too well where we can expect to be led by the nihilism entailed in rushing in chorus into the fog. Absent a sense of the big picture, a movement is blind. Without a doctrine (but I would rather say a *Weltanschauung*) to throw light on the panorama – human freedom and the earth, ecology runs the risk of being paralysed, if not torn apart, between its left and its right; or at least, of keeping to a superficial confusion that will leave it defenceless before the opponent, which will hijack it sooner or later.

This very label of ecology, borrowed from a specialized science, shows how incomplete is that scrutiny, for example, when it comes to the role of science in development. For many reasons, reflection is needed, even while it is barely starting. The ecological movement is barely ten years old, whereas socialism has had over a century to debate and clarify its endeavour. And if socialism put into question the economic structure of a bourgeois society that was a lot more primitive than ours, it was in the name of its very values of science, education, work and progress. By contrast, ecology tackles a social organization that is infinitely more complex, that encompasses everything and whose repercussions are of every kind, not only economic and social but biological as much as intellectual and spiritual. It is truly a world it takes on since the planet is at stake and, what is more, it changes every second. Certainly, never in history have humans thus been compelled over so short a time to effect such a revolution of the whole; and so, to some extent, in the values of the times.

If we want to understand and master the countless manifestations of 'development', we do have to start at the root by elucidating the reasons it invokes in light of our own. In a new context, we will not be able to elude the fundamental questions that have always haunted and determined man, failing which we will get lost in the present jungle, without being able to make out a common direction. Namely: What is nature and what is man? Where do we come from and where are we going? (It is mostly the second point that matters to this writer.) We are not about to finish, since a Marx or a Reich has not brought us the innate ecological truth. Above all in a society in accelerated motion, which forces us to take our bearings at every moment. If, pressed by urgency, we cannot stop being glued to current affairs, we will fail, either out of an inability to judge the situation for lack of distance or, worse, because we have succeeded at the price of forgetting our reasons for being. Our paradox is that of having to hurry slowly, being commanded to reflect in a house that is on fire. But we can no more force the development of a great human change than that of a plant.

One can build only on a solid foundation if one is clear about oneself, one's situation and what one means to do with it. And this fundamental research is needed for action for another reason. Failing a nuclear disaster or the like that would turn public opinion around, the ecological movement must count on itself. The power of a fledgling movement resides not so much in circumstances (number, material power and influence over the public) as in itself: in the vigour and quality of its supporters' spiritual, mortal and rational motivations, without which such an endeavour would lack an engine. A movement is above all rooted in its members' personal convictions. The more they are, the more the movement will be solid and alive, because it draws its sap through countless and deep rootlets. If for each one the motive for acting is truly a matter of life and death for which the chances of success are secondary, then he will be resilient in the face of failures and the wear and tear of time. And it is conviction that sparks imagination, the faculty whereby, by dint of hitting one's head against the wall, one discovers or invents the crack that makes the impossible possible. And if by some mischance a serious crisis erupts, which could very well happen in the present chaos, it is the force of convictions that will enable one to hold on amid the storm by facing solitude and the risks of clandestine action. Each member's heart of hearts is a movement's final stronghold in the event of unrest. By contrast, someone who identifies with his organization, like German social-democracy in 1933, dissipates like a ghost at the first impact.

The energy of personal motivations – living faith, the antithesis of ideologies or mythologies that are but cement to stop emptiness – can alone bring men to federate with others in spite of the diversity of their religious, professional or national origins. The more a certainty is alive and deeply rooted, the less it has need of ready-made ideological armour, truths and a framework to defend itself and the less it fears criticism and closes itself to others. It alone can inspire what is neither a dogma nor an organization, but an alliance. Realism compels one to acknowledge that the ecological movement brings together participants who differ sharply in temper, age, milieu and concerns. That is its interest and its difficulty. If each one does not come out of his speciality or of his small corner, ecological action will scatter in different or parallel and competing institutions. This divergence will be avoided only if we are well aware of the unity of the industrial phenomenon and of the reasons that impel us to fight it. This while being perfectly clear about what distinguishes us, so as to be clearer – without lies or illusions this time – about what unites us. An association where agreement emerges from differences is a great and difficult novelty, but a world that threatens them all drives us to it. Let us recall that there is no stronger motive that can bring together the most diverse people than the threat of a total destruction of nature and freedom. It is up to us to ceaselessly think of it, even if it is sometimes hard to look straight into this black sun.

Weak points of ecological thinking

The ecological movement has lacked the time for maturation. Before 1970, the description and the critique of 'development' were limited to isolated and scattered writers for whom writing was an end and who did not feel the need to compare their experience with that of others in view of some kind of action. Either that or the denunciation of the rape of nature was the province of ecologists (this time in the scientific sense of the word) given to extrapolating its laws to the human realm, content, as far as social and political facts were concerned, to complement this specialized viewpoint with ideologies of the Right and Left borrowed from the very society they were denouncing. As for the ecological movement, born after May 1968, along with the green light of 1970, too often it too has been content to complement ecology with an anarchistic ideology dating from an era that did not yet have to worry about the industrial system's totalitarian tendencies. This is understandable; born too late, it was caught up in emergency, in a world that despises reflection, that shoves it around by multiplying information and obligations, particularly among bona fide intellectuals whose agenda is already full.

Ecological thought has thus lacked the contacts and the discussions between people that would have allowed it to throw light on its own contradictions and to overcome them, and to round out a thinking that often remains inchoate whenever it comes to man.

To be sure, if we take the work of the various authors who have been raising the issue for a long time, rather than that of the eleventh-hour defectors, we find in it all the elements of a coherent critique. But these works are scattered, and due to the craze for topicality, ecologists hardly know them. This is why their program is more of a ragbag of critiques and solutions in which the essentials are mixed with the accidentals and where gadgets hide the true remedy. Looking carefully, we realize that as an aggregate, everything or almost everything has been said, albeit without any sense of priorities. And the effect of the detail hides the encompassing dimension that determines it. From this point of view, the ecological movement has the opposite fault as Marxism which, holding the one Explanation, is not afraid of practicing the first-of-all and second-of-all and the 'make sure you stick that into your skull'.

For the average ecologist and the public, ecology is above all the refusal of peaceful nuclear power. Let us add to it, far behind, the campaign against the Larzac camp and chemical pollution, the campaign for the development of solar power and, lesser known still, that for organic agriculture. Everything else is of concern only to isolated individuals or defence committees that do not go beyond their local horizon.

Hence some significant weak points. The place granted to antinuclear struggle, which is indeed of prime importance, is liable to obscure the far clearer threat of an apocalypse unleashed by the spread of nuclear weapons. The supporters of atomic energy are not entirely wrong when they claim that the refusal of peaceful nuclear power is dictated by the suppression of anxiety about nuclear war. Peaceful nuclear energy entails the risk of serious accidents, while the spread of the ultimate weapon is the certainty of apocalypse: no need of experts to teach us this. But it is better not to think that death haunts our house and to attach our fear to a risk of the same nature, only less certain. This by the way leads to an overestimation of the dangers, which are already great enough. The ecological movement suffers from a catastrophism it inherited from Fournier who, as far as he was concerned, knew that the end was near. In spite of oil spills and nuclear leaks, the 'pedagogy of catastrophes' has yet to work, to the great satisfaction of the pronuclear lobby that uses it as an argument to underline the reliability of this technique. Which is why the real catastrophe, less spectacular, is going to happen. For it will be avoided only at the price of a reinforcement of the secrecy and police control that alone can prevent an accident or a terrorist act. Catastrophism is right to announce apocalypse for tomorrow, but this tomorrow is not that of journalistic sensationalism, it is that of the history and species that are coming, step by step, adding up in silence pollution and restrictions on freedom. It is only today that we learn that the car means millions of dead and wounded. Suppose we had learnt it all at once in our morning paper, what a catastrophe! The true catastrophes are only ever known in the long run, when the damages are irreversible. This may be the case of oil and the endless amount of garbage that accumulates in the oceans on a daily basis.

As a matter of fact, we are living in an explosion; if it does not appear to us as such, it is because the human life span is shorter still. The true catastrophe is development. And we must not forget that it is still going on. More ... always more ... accidents and crises, and especially garbage of all kinds and constraints. Development is the taboo a priori, which does not even need to be proclaimed, relative to which everything is organized against the grain of the rising shortages it generates. What is the growth rate going to be?

Is it 7, 4 or 0.5 per cent? It does not matter; until the last moment, everything is going to be sacrificed to ensure that this percentage is at least one of growth. Development, and so the necessity of curbing it, should be the *delenda est Carthago*[1] of which we are reminded every moment, failing which the various critiques or actions of the ecological movement lack a foundation. If it wants to designate an adversary, nuclear power only shows one aspect of it, while development includes them all. By shining the spotlight on one particular sector, everything else is left in the dark, and the enemy seizes the opportunity to extend his grasp. And a locally positive solution turns against you. By choosing the train against the car without questioning development, one ends up supporting the TGV, and so one more step towards the destruction of Burgundy. Whereas if we put the real question: Why Paris–Lyon in two hours instead of four? the answer becomes very clear. Solar energy itself, within the framework of development, from a gentle technology, becomes a hard one by covering an entire *canton* with a funereal shroud. And of course, this technique, being concentrated to whatever degree, will fall into the hands of corporations or the State. Actually, it is already starting to.

The countless threats that mobilize ecologists and risk scattering them are but the multiple effects of a single cause: the religion – hence the blind practice – of technical and economic development (or growth). This is the central point that everything else comes down to, be it nuclear power, the disappearance of forests, the paving over of coasts or pollution and so on. We will not be able to do anything if we refuse to question the myth of an indefinite development of production and profit, of the multiplication of jobs through the exponential progress of industry. It is on this point that it will be possible to distinguish the true defenders of nature and man from the official zealots of 'the environment' and 'quality of life' who think they can produce nature reserves and national parks thanks to petrochemical complexes.

The scant attention paid, except maybe by the very bourgeois Club of Rome, to the cause of all effects no doubt explains the absence of a reflection on the drowning of space by the economic

---

[1] Translator's note: This Latin phrase for 'Carthage must be destroyed' is associated with the Roman senator Cato the Elder, who is said to have used it as the conclusion to all his speeches.

deluge. Even though its rarefaction due to industry, infrastructure, the military or tourism encompasses all other shortages: of energy, of raw materials, of manoeuvring grounds, of sunny beaches or of skiable snow. The failure to tackle the Phenomenon from the angle of space explains the scant attention paid to its biggest consumer, tourism, whose explosive development now extends all the way to the valleys of Nepal and the atolls of the Maldives. Why such an exodus and its modalities? In what way does it answer a deep human need, unless it is merely the product of the progress of cars and planes, of fashion and the advertising of tour operators? It would be in the ecological movement's interest to ask itself these questions – and others – if it does not want to leave them to a sociology financed by the Ministry of Tourism and Club Med. It should not feel intimidated by this pseudoscience; its attachment to nature and freedom will allow it to push objectivity much further than Research for wages, which is compelled to spare, or even supply with justifications, the powers financing it.

The absence of reflection on the growing shortage of space also accounts for the fact that next to nothing has been said on the problem of land-use planning made necessary by the rarefaction of its object, which it will only save by trapping it within the iron shackles of a POS[2] or urban planning. I will not get into the details of this destruction of space – that is not exactly the topic of this book.[3] But it is safe to say that, as long as ecology does not attack industrial society at this level, its critique is going to remain piecemeal.

Another 'detail' on which reflection remains almost entirely missing is science. This even though it is an essential point of the critique of the industrial system since, in the ruins of religious and moral dogmas, science supplies it with universally acknowledged truths.[4] A radical critique hits at the root: as yesterday about

---

[2]Translator's note: Public Open Space, English acronym in the original.
[3]Translator's note: Charbonneau would partly devote a posthumous book to the growing dearth of space and time: *Finis terrae* ('Land's End') (La Bache: À plus d'un titre, 2010).
[4]Translator's note: Hence the Latin title, meaning 'the final argument', of Charbonneau's later book on the topic: *Ultima ratio*, paired with the other side of the coin: *Le Paradoxe de la culture* ('the paradox of culture' as gratuitous entertainment serving to provide hard-nosed science with an alibi) in the diptych joining the new work to the latter reissue of a previously published book (Paris: Denoël 1965) under

religion, it must raise questions about science, even if this means bowing to it where its authority proves legitimate.

What is science, what are its place, its gains and its costs from the standpoint of nature and especially of freedom? The question has hardly been raised except by the mathematicians of the group *Survive et vivre*. And the critique (or problematization) of it that has been carried out for a long time already by a few mavericks, strangers to the dogmas of progress as much as to those of the church, for example, the questioning by Nietzsche as an expert in ambiguity, has hardly influenced ecologists. Like our society, they keep swinging between admiration and fear of science, viewed as a kind of magic that can be white or black.

Since the movement includes many young people and dropouts who have only a vague idea of scientific research, it mixes scientism with anti-scientism and thinks it can find the solution for everything either in ideologies drawn from a specialized science such as systems analysis, or most often in parallel sciences that lack the rigour of mainstream science. Such a return to magic can only discredit critique and justify the imperialism of a specialized knowledge that will end up invading everything, at the risk of getting lost itself, leaving nothing for personal and popular knowledge – and hence power. Furthermore, this absence of critical spirit leaves it defenceless before the rise of new sciences whose effects are likely to be even more severe than those of nuclear physics. This is, for instance, the case of biology and genetic engineering. It would be unfortunate if the mythology of everything 'bio' led ecologists to believe that biology is more innocuous than physical and chemical sciences. The manufacturing of new bacteria in labs or the careless spreading of exotic algae, plants or predators can break equilibriums; and the use of insects under the pretext of 'integrated control' is likely to be even more disastrous than that of pesticides.

The absence of critical reflection on science explains the scant attention paid by ecologists to a major scientific mutation: the emergence of (so-called) human sciences as the condition for techniques of manipulation of humans in the same way that natural sciences are the condition for techniques of manipulation of nature. All kinds of questions come to mind for one who is concerned.

---

the common title *Nuit et jour* ('Night and Day'; Paris: Economica, 1991, 'Classiques des sciences sociales' series, ed. Hervé Coutau-Bégarie).

Are they really sciences? Can (and should) man be an object for man? Who can really know him except a god? And to treat him as a thing, would it not be first necessary to negate him or, if need be, to annihilate him? Which laboratory is going to take it upon itself to manufacture persons and society, and in the name of which criteria? Will the science of man not be, like religion in older times, merely the mask of power? And if it is truly scientific, would this not be worst of all, insofar as it will supply said power with efficient weapons? Now that the sciences of man, still in their infancy, are taking their first steps, this is the moment to become aware of it. If, after nature, man becomes the plaything of scientists, what share of knowledge is going to be left to individuals and the mass of the people? Scientific imperialism having invaded the other hemisphere, invisible until now, of human existence, where will there be an outlet for the need to think and to create of someone who has not received the 'research' label? And when he leaves his specialty, every scientist is this man. And lastly, will not science itself, by touching this realm, finish giving itself over to the Market or the State for which control over men is even more vital than control over things? As the year two thousand dawns, these questions are crucial.

Science is essential, for without it, technique and organization would not exist. Hence the grants and the favours from corporations and the State: the time of the 'great distress of laboratories' is well and truly over. We may fuss over the distinction between science and technique, but the link between them is indisputable. And everything else flows from it. Here too, ecological reflection has its work cut out. It has sensed a major self-evidence that only a few here and there had recognized, namely that technique is key in an industrial society it fashions by making it ever more massive and centralized or, in other words, organized. Hence the correct idea that if we want to change life, it is not enough to blame the economy; we must also change technique by adopting gentle, decentralized techniques. Let us not expect them to be so on their own; that such and such a technique – cable TV for instance – will automatically give us freedom. Any technique – power – is hard to a lesser or greater degree. It is the human caution and will not to abuse it, imagination, which makes it gentle. A good use of science and technique assumes a Copernican reversal: creating the means of one's society rather than the society of one's means. This will not be done in one day thanks to the miracle solution of solar energy; especially if one is advocating *small is beautiful*[5] and gentle techniques without refusing development. To what extent is the science of necessity not tied to power over nature and men, to technique? In what way does the latter, the source of a power that is indispensable to freedom itself, bring with it a division, and hence an abstraction, a growing complexity and centralization of work and life? We must at least raise these questions if we do not want scientific and technical means to become the end that commands everything.

Technique leads to organization, whose supreme expression is the modern State, the organizer of organizations. Can it be avoided, and if it cannot, how can it be prevented from becoming total? We should not confuse the dreamed-up solutions that dispense one from raising the question with the critique and the struggles – ever to be taken up anew – that will limit the field of organization by reserving that of nature and freedom. The scant attention devoted to organization no doubt explains the lack of attention to the disorder that everywhere follows rationalization: thus, the decadence of certain services such as the SNCF,[6] and especially the PTT,[7] sometimes rife with tragic consequences for users.

Ecology has not concerned itself enough with techniques which, by facilitating concentration like IT, threaten freedoms: all too often, it has been content to incriminate the big bad capitalist or governmental wolf. There again, the issue remains vague. To what extent do these techniques remain masterable, understandable and usable by anyone? The same remark should be made about the media, whose critique has hardly been promoted by ecology, while its action, out of a concern for reaching the public, has largely been determined in view of the platform they offered it.[8] There again, is it not paying too a high a price in various servitudes for the limited

---

[5]Translator's note: In English in the original, from the title of the influential book by E. F. Schumacher: *Small Is Beautiful. A Study of Economics as if People Mattered* (London: Blond & Briggs, 1973).
[6]Translator's note: Acronym for Société Nationale des Chemins de Fer, France's State-owned railway system.
[7]Translator's note: Acronym for Postes, Télégraphes et Téléphones, France's combined State-owned postal and telecommunications service, until the latter were separated from La Poste and eventually privatized by the end of the twentieth century.
[8]Translator's note: Charbonneau would develop this critical media ecology in a self-published book, *La société médiatisée* (1987).

audience they provide? Is it possible to alter them? Until now, in spite of pirate radios and cable TV, the answer is not very encouraging for freedom. And one would at least have to raise the issue.

Shortcomings in economic and especially social reflection

The critique and the curbing of development too often remain in the background, which prevents the search for and the definition of an alternative economic policy. If, carried by mobilization against nuclear power, ecological reflection is fairly advanced on energy issues, it displays some serious shortcomings in other areas.

The May revolt instinctively attacked one of the foundations of bourgeois or socialist industrial society: the myth of work. But as with self-management, the 'against' retorts to the 'for' without prior scrutiny of the practical and psychological necessity of work that acutely manifests in all societies and above all in communities. To what extent can we escape this curse and its double: unemployment? Is it not rooted in man who, very often, draws his joys from its troubles? Could he do without it? Where does labour end and that work begin wherein effort becomes a joy? And from the moment it becomes collective, does it not include ever more pure labour, divided and abstract? Hence the compensation of leisure in our society, which tends – except for some leaders, researchers or artists – to tear apart life between a work time that is reduced because it is concentrated and a 'vacation' time devoted to pleasure and freedom in nature. Instead of work, is not the worst curse that of this dichotomy that decomposes our time as it does space between leisure areas and industrial areas? And is not the true solution the reintegration of nature and freedom in work, and of the seriousness of work in the frivolousness and relaxed attitude of leisure? Ecological reflection is only beginning, even as it will never be through.[9]

The economic obsession that characterizes the 'people in charge' in our society risks bringing about among its opponents a negation

---

[9] See Bernard Charbonneau, *Dimanche et lundi* ['Sunday and Monday', Paris:] Denoël, 1964.

of the economy that confines them within discourse and spectacle, whereas it is now a matter of directing it towards ends other than itself: the preservation of earth and freedoms. He who neither forgets ends nor the weight of means cannot but be crushed by the enormity of the problems raised by such a shift. How might it be possible to make people accept such an about-turn? And this being done, how can we manage it without causing a machine organized around a wholly different principle to explode? We will no doubt never have enough firmness about ends and empiricism about means. It is true that the protection of nature, along with human variety and freedom, will be ensured only if we dissociate the national or multinational economy into small autarchic or self-managed units. But then we fall back into the problem of local societies and their federations. How do we ensure economic unity in diversity in a form that is neither that of the capitalist market nor that of the State plan? To get over the obstacle, we must first size it up.

In economic matters, ecologists are too often content to take up a left-wing critique of capitalism that has given many proofs of its unrealistic or even nefarious character. And some groups have welcomed as scientific truth the miracle solutions of another era, such as Duboin's theories on abundance,[10] while the business of our own time is to put an end to the shortages it brings about. This is why we are still in the fog when it comes to money, capital, currency and finances, eliminated once and for all – at least in discourse. There is however good reason to ask ourselves some questions if we do not want them to reappear – only twice as bad – after their suppression, as in the USSR. The condemnation of 'consumer society' (which of course remains one of production) has been a little hasty. Which capitalism? Which money? Is it Harpagon's gold[11] or today's melting money? Which one possesses you the most, the one you have or the one

---

[10]Translator's note: Jacques Duboin (1878–1976) was a French banker and industrialist who left politics to develop and spread his own theory of 'distributive economics' in publications that had some marginal public impact in the 1930s. Sometimes called *abondancisme*, it argued that the plentiful production of the machine age made obsolete the economics of scarcity, typical of rural societies but still assumed by the received wisdom of industrial ones.

[11]Translator's note: In Molière's classic comedy *L'Avare* ('The Miser').

you constantly must be running after? The same thing goes for property: everyone denounces it, but each one dreams in secret of having his own dacha. Which property? The one you love because you take care of it and inhabit it, and where you retire far from the roar of the social ocean? Or the abstract one, indefinitely capitalizable, from which one expropriates others and gains power over them? Can there be freedom without property, be it individual, the family's or the community's?[12] These questions at least deserve to be raised.

And there remain quite a few to sort out in an essential sector: agriculture. Ecologists dream of returning to the land, which they confuse with free nature. They love its charms and vigorously denounce the ravages of land consolidation, pesticides and fertilizers. But they are far from grasping the full gravity of its destruction and hardly see any solution other than this one: tautologically termed 'organic' agriculture, as if any genuine agriculture was not a technique of living matter. This issue takes a back seat, while the current transformation of agriculture into a mechanical and chemical industry and of the countryside into an agribusiness suburb is without a doubt a change even more major than nuclear power. Let us rehash its costs under the guise of slogans. (1) Since the countryside represents 90 per cent of France's space, the industrialization of agriculture extends the effects of industry (pollution, landscape destruction, etc.) to 100 per cent of her territory, including the few national squares, which it turns into a total suburb from which it is impossible to get out. (2) By the same token, countryfolk and city dwellers are deprived of the tasty and varied foods that turn meals into a daily pleasure. And the agribusiness industry will take care of lovingly cooking in its factories those insipid lipids, carbohydrates or proteins. (3) Finally, the desertification of countrysides, the standardization of work and that of farmers' leisure by television annihilates the villages, cantons and provinces that made the cultural richness of European nations. Nothing to see, to eat, to meet: no point in getting out of one's

---

[12]I have treated these two topics in two unpublished essays: *Il court, il court, le fric* [Bordeaux: Opales, 2005] and *La propriété, c'est l'envol*. Translator's note: Both titles are untranslatable plays on French cultural references, the former a nursery rhyme title applied here to 'dough' and the latter a positive spin on Proudhon's famous dictum on property as theft.

car. Here, an atom bomb is in the process of exploding. Under my window, walls are falling, the earth is opening and the black shards of the ONF[13] plantations stain what was once the bronze flank of a tree-covered moor.

To answer such a threat, organic agriculture is not enough. Here again, is it a matter of feeding the dream of a purely natural life or of foiling the imperialism of the total ersatz? The minimum would be to compel the latter to own up to what it is: a branch of mechanical and chemical industry; and organic agriculture is hardly forcing it to do this by calling it 'classical agriculture' when it has nothing to do with agriculture, and a classical one at that. This merely justifies the claim of corporations and technocrats to defend the agriculture and peasants they are busy liquidating. What is agriculture? Up to what point can it deserve its name by transforming nature and its fruits? Doesn't the mythology of the purely natural product lock up organic agriculture in a ghetto of the luxury artisanal product that can be perfectly integrated within the market of agrochemical food plastics? The debate has not even begun.

In this area, the ecological opposition cannot even manage to get away from a system that encompasses it. It is not surprising that an industrial society hungry for space and outlets blocks out the agricultural question and hence that of food and cooking that directly depends upon it. Among other considerable mutations, we must point out the setting up of an agribusiness industry (canned food, frozen meals, etc.) run by corporations with the help of the State's laws and subsidies. Which means that the French are going to finance the end of their family polycooking. Here again, freedom is being destroyed in the name of freedom: of the emancipation of woman, who is no longer chained to her ovens thanks to the *ready-made*.[14] Be it for good or ill, it is without a doubt a fundamental change. For food (as opposed to feeding cattle) nourishes the spirit as well as the body and has always allowed company to gather as dinner guests around the Holy Table. *Der Mensch ist was er isst*,[15] a man fed with ersatz becomes an automaton himself, and the society

---

[13]Translator's note: Acronym for the Office national des forêts, France's National Forest Office.
[14]Translator's note: In English in the original.
[15]Translator's note: 'Man is what he eats' is a famous quote from the German Left Hegelian philosopher Ludwig Feuerbach (1804–1972).

that allows this last holdout of cultures to be invaded is no longer even a shadow of its former self.[16]

But as it happens, on this point ecology is pretty much useless. Vegetarian or not, it keeps on unwittingly obeying the old Christian taboo that condemns the sin of the belly. Food is unclean, especially meat and wine, and its delights are stolen from the Third World. Following Dumont and a dietary science that would like nothing so much as to put Frenchmen on an industrial ersatz diet, it denounces the Grande Bouffe.[17] These ecologists have a point, only they forget that the quantitative abundance ensured by industrialization is paid with a dearth of quality: of tasty and varied foods; and above all for the less wealthy who cannot afford kind-of-farmhouse cheese at La Maison de Fleurance. Their critique of agribusiness remains that of the Left: that of capitalism and multinationals (see Garreau's *Agrobusiness*[18]). This is one aspect of the question, but a secondary one compared to the concrete change brought about in daily life by the transformation of nutrition into a science and an industrial technique. As elsewhere, this left-wing critique obsessed by the capitalist producer forgets the main thing: what he produces. Will the textured soy protein or the urea-soaked petroleum steak[19] listed on our menu be capitalist or socialist? Will the State that will force-feed it to the people have a red or a pink flag? That is the crux of the issue.

The ecological movement's concern with concrete facts has not extended to the area of nutrition. With a few variations, it is doing no more than repeating an official discourse common to left-wing

---

[16]See on this topic *Un festin pour Tantale* [Paris: Le Sang de la Terre, 1997, 2012], which has yet to be given the green light (Bernard Charbonneau).

[17]Translator's note: *La Grande Bouffe* is the title of a 1973 cult movie by Marco Ferreri about a group of French bourgeois who deliberately eat themselves to death as part of an orgy in a country villa.

[18]Translator's note: Gérard Garreau, *L'Agrobusiness* ([Paris:] Calmann-Lévy, 1977).

[19]Translator's note: A process for the extraction of proteins from hydrocarbons, using urea to separate out the paraffin, was developed by the French branch of British Petroleum (BP) in the 1960s. When industrial production of POUP (*sic*: *Protéines Unicellulaires d'Origine Pétrolière*) started in plants in southern France and Italy in the early seventies, it was hailed and promoted for its potential for mass animal and human consumption by prominent scientists, nutritionists and Third World experts such as Green presidential candidate René Dumont. However, this fledgling industry and the process on which it was based were soon suppressed by BP under pressure from the US soy lobby.

Christians and agribusiness CEOs. Clearly, we will not escape the single dish, with seconds and more seconds until we throw up, and a zest of organic lemon on Sundays. For if the socialist revolution has shown its ability to feed the people by imposing rationing, neocapitalism, on its part, rationalizes, even if it covers up the disappearance of varieties under all kinds of chemical colours.

As for society proper, the more we advance into the human realm, from the economic to the social and from the moral to the religious, the more ecological reflection is absent or borrowed from other sources. Breaking with ideologies inherited from the pre-industrial era, we can at most mention Illich's critique of the power of medicine and schools, previously pretty much a taboo. It takes up the critique developed by marginal reactionary or anarchistic theorists from the early twentieth century. But there too, things are immediately pushed to the limit by negating, instead of qualifying, the authority of medicine and schools, which is hardly conducive to resisting its abuses. There is no point in denying authority to doctors if it is to grant it to healers.

One would never end if one were to point out all the weak points due to the absence of an original ecological reflection. Since René Girard is not in fashion yet, we have seen that ecology has hardly made any strides in the critique of violence and war. As the mortal threat posed by the spread of atomic weapons recedes in the background, so does its cause, nationalism, which anti-Americanism or anti-Sovietism leads many to treat with some indulgence. And yet the myth of national independence, perforce a deceptive one for small and middling States, leads not only to a justification of the proliferation of 'strike forces',[20] but to economic and demographic development to keep up with the competition of neighbours: the economy is a war whose challenge must be taken up. At every moment, ecology meets nationalism on its path, whether it is destroying local nature in the name of the nation or it refuses European or world legislation protecting the planet in the name of the sovereignty of France or of Burundi. If ecologists want all ethnic groups of the world to someday federate their differences so as to save them together, they would do well not to identify them

---

[20]Translator's note: In the original, *'forces de frappe'*, adding a plural to the official name of France's nuclear arms doctrine.

with national particularisms which, by the same means, absurdly pursue the same goal: political, military and industrial power.

There remain morals and religion, where ecology can only fall into the wide blue yonder, which is at once white, red and black. We have raised earlier the question of sex and the family. Can we speak of natural differences or institutions? Is it possible to conserve or to transform them? If so, which ones are going to be replacing the old ones? The commune or the State? Unless the return to nature is that to pure freedom, which I doubt. In any case, this is something to contemplate while waiting for the next vote. But of course, one can also settle these pointless questions by refraining from asking them of oneself. Which is what has been done under many guises for two thousand years.

If, turning our back on current affairs, we wonder about this part of ourselves – society – before coming back to the present by different paths, we finally hit against the question of the meaning of life and action that man is no doubt alone of all nature to raise. But most often, religion is not a question, but an answer home-delivered by the local guru. Not being from the firm, I do not need to sell this tranquillizer to ecologists; I can at most throw light on their path. This is why I have placed there the two signs I took out of the public dump: nature and, above all, freedom.

CHAPTER NINE

Recycling

Ecology is now all over the place. With the first signs of the energy crisis and of the crisis of Progress, it invades the speeches of the media and the ministers, which does not necessarily mean that they translate into acts. A solitary realization becomes a social fact signified on TV. But by the same token, is not the ecological movement reabsorbed into the society it purports to transform?

Creation or recyclable social by-product?

Every society defends itself, and ours has at its disposal more powerful means than any other. It represses by breaking, or more simply by grinning and bearing. Above all, it reuses or recycles any social refuse harmful to the body politic; in high doses this would kill it, but swallowed in the right way at the right time, in the prescribed dose, it perks it up. This happens spontaneously, that is to say by *feed-back*,[1] in every possible way: the first industrial society had diverted the demand for nature into bucolic literature and tourism, the second one adds to the latter technocratic recycling and leftist entertainment. Having first carefully sorted what could be assimilated and what was harmful – there could not be any compromise on that – it recycles for its own profit part of the opponents' ideas, co-opting by the same token the bulk of their troops and of their leaders, seduced by the efficiency and

---

[1] Translator's note: In English in the original.

the advantages of an official career. The established order also defends itself by marginalizing, that is, by driving back the threat far from the centre in out-of-the-way regions or areas; or better yet, to the antipodes of the real: in luxury, poetry and the quality of life, which hardly exists without quantity. This is possible only with the complicity of the enemy, who allows himself to be locked up in harmless extravagance; that of the extremist or, better yet, that of some generously waged professional of spectacle, which allows the public to release the pent-up impulses repressed by the social order.

Wherever a small kernel, being too hard, cannot be broken nor assimilated, society encysts, somewhere in a corner, a ghetto beyond which an ideology or a sect no longer threatens to contaminate the whole of society. Here again, this sectorialization process can only function with the opponent's participation. This is how the bourgeois world has been able to contain communism with the complicity of a giant sect devoted to Moscow. Either that or society depoliticizes by disconnecting from the whole the young diehards it causes to retreat into abandoned countrysides where, after a time of exile, they are purged of their illusions about living free and are ready to be recycled. Or else society depoliticizes by getting the movement that threatens it involved in an electoral campaign that becomes its only end: as much as the representation of social currents and their participation in power, the function of democratic elections is to digest conflicts without violence. And if, by some unlikely chance, all these defences fail, there remains the ultimate means: revolution, where society, like the Phoenix, rises from its ashes. Attaining power can be the worst of defeats for a genuine opposition. The revolution of 1789, and that of 1917 even more, have only destroyed society by regenerating it: by effecting a strengthening of government, the economy, schools, the army and the police that the old regime had not been able to achieve. Having as its calling a much bigger change, without which the earth and the species may well be lost, the ecological movement cannot afford the luxury of these various forms of failure.

This, without even admitting it to itself, industrial society knows, and it works at eliminating its opponent by any means available. If the ecological movement, solidly rooted in its ends, is not on its guard, it will suffer the fate of its predecessors.

Recycling through fashion and fashionistas

Being the fruit of a deep personal and collective reaction to the upheaval of the human universe, the ecological social fact is also a fashion. It displays all its features: a sudden appearance, stereotypes and shiny gadgets, spectacular repercussions through the centre's media, recycling by commerce and politics. If ecologists are not careful, their movement may well suffer the falling out of fashion that inevitably follows being in fashion. Since 1975 we can feel it coming, but the issues raised by the rise of industry are too deeply seated in social life to be completely forgotten.

Just as tradition characterizes relatively immobile old societies, fashion is the hallmark of a society like ours, in perpetual motion. There is another, antithetical reason. Since movement forces us to hang on, and our dynamic societies impose even stricter models than the old ones, change fosters the dream of change: of an elsewhere, cheaply satisfied by tourism and, above all, by spectacle. Hence the dizzying acceleration of fashion's merry-go-round, condemned to an eternal return drawing from the limited stock of forms and formulas of the past that it recycles after a certain time of forgetting.

This function is ensured by a vanguard of fashionistas: intellectuals, artists and actors who are specialized in this formal production and tied to the media that vulgarize it. Some are manufacturers of ideas, others of more or less new forms that, automatically abstracted from reality, the realm of economics and politics, are now kept in a cultural ghetto. This way, society makes sure they remain harmless at least in the short term; it does not care about the long one. This entertaining vanguard (although, to tell the truth, it belongs to the army just as much as the clique that precedes it in magazines) is of the Left. Whereas the Right, silent as the very way of things, reserves for itself economic and political power, abandoning intellectual power to scientists or entertainers who do not have the means of exploiting it in practice. Society thus vents in a kind of noisy but harmless explosion the energy that might blow it up. Understandably, this position, at once necessary and pathetic, is generously paid.

The specialized intelligentsia thus has as its mission statement the search for ideas and new forms. It creates fashion, or gets hold of the one that spontaneously arises among the youth it must follow with great attention. And this attention is all the greater as competition is even sharper among the intelligentsia and within art as it is at the grocery, which it supplies with enticing packaging whose model often comes from the United States. This explains how promptly ecology was taken up. This is not a matter of conversion to ideas – all are good when they come in their time – but of adaptation. The intelligentsia, particularly in the intellectual sector proper, has become ecological to various degrees because, in order to remain at the vanguard and in touch with youth, a touch of green was needed. It is the same sociologists who have been in turn Stalinistic, then de-Stalinized and Americanized, who have now taken up ecological themes. No need to give names, these 'conversions' have nothing personal about them; in some sectors all have undergone them. This is particularly the case of agronomics: all the fathers of agrochemicals are now ecologists; none is missing now that E. Pisani[2] has joined his colleague Mansholt[3] in the defence of the environment and polyculture. Everyone is in favour, including the sociologist of the development of hybrids. The INRA[4] too, which is all smiles for *Nature et Progrès*, while the latter feels flattered to have relations with such a high scientific authority. But land consolidation, landscaped or not, proceeds apace. The case of the Club of Rome, a synarchy of illustrious technocrats (among others for France the former directors of Datar and the Plan) may be different.[5] Since we are now talking about men positioned at the

---

[2]Translator's note: Edgard Pisani (1918–2016) was France's second longest-serving minister of agriculture, reforming it in the 1960s to foster a productivist and export-centred agriculture in the country, and later in the European Common Market as an architect of the Common Agricultural Policy.

[3]Deferentially interviewed in *Nature et Progrès*, the review of organic agriculture, the man responsible for the Mansholt Plan (who remembers it?) has answered without shame: 'I have not signed a lease with truth.' There is a time for everything: to chase away farmers, and then to make them come back.

[4]Translator's note: According to the website of the French National Institute for Agricultural Research, 'INRA is Europe's top agricultural research institute and the world's number two center for the agricultural sciences. Its scientists are working towards solutions for society's major challenges.' http://institut.inra.fr/en.

[5]Translator's note: From its origins in Vichy regime propaganda, the term 'synarchy' for an international technocratic cabal has remained a mainstay of French conspiracy theories of all stripes. When it was created in 1963, DATAR was the name of the Délégation à l'aménagement du territoire et à l'action régionale ('Land Development and Regional Action Delegation'); in its current incarnation, the second A now stands for '*attractivité*'. In charge of economic planning (e.g. five-year plans), the

actual wheel of the system, we can always tell ourselves that Aurelio Peccei was well-placed to have an idea of what the end result of his activities would be.[6]

Were it not for the rapid succession of its turns, there would be nothing new, hence nothing to wonder at about this opportunism. On the other hand, the ease with which the ecological movement has allowed itself to be colonized by professional recyclers is more surprising. This may raise concerns about its future. This being said, it has no shortage of excuses. It is young, it too having been born of current affairs in May 1968. Hence its ignorance about its true forebears, such as Roger Heim, eclipsed by Illich and Dumont. As for the current fathers of the ecological movement, they are its old children: worthies who have access to the media and are running after youth. Without them, how would it be possible to get known in the press and on TV and thus reach the general public? When you are young and feel self-doubt under all your bluster, how can you fail to feel flattered to be received by famous people? All the more so as, for some social climbers, ecology was their age group's footboard of choice. And the bourgeoisie is always ready to open its arms to them whenever they get tired of barking for nothing on the extreme Left. Under the fond gaze of daddy president, they are finally acting.

This has nothing to do with morals. It all fits: by granting authority to recyclers, the ecological movement runs the risk of getting recycled in its turn. Instead of knowing itself, it turns to mainstream sociology and, in 1979, Alain Touraine informs it of what it is: namely, that there is an external conflict between ecologists and technocrats, and an internal one between its 'environmentalist', 'political' and anarchistic strands. Something it might have discovered on its own. Why turn to a specialist who has demonstrated on many occasions, like his colleagues, his ability to follow, and not forecast, a context successively placed under the sign of Stalin, American science and finally ecology? One fails to see why an expert who has twice been wrong about the nature of society would not be wrong again a third time. Opportunism never records

---

Commissariat général du Plan existed from 1947 to 2006, and now goes under the name of France Stratégie.
[6]Translator's note: This Italian industrialist had rebuilt FIAT, founded Alitalia and become president of Olivetti by the time he launched the Club of Rome in 1968.

what will be, nor even what is, but only what was. The ecological movement has nothing to learn from the reigning intelligentsia, which has given many proofs of its scant intellectual autonomy. It is hard to imagine a nascent socialism seeking the counsel of the philosophers of the Sorbonne. The knowledge of ecological action is going to be the work of ecologists themselves.

Recycling through technostructure

Through the voice of its highest scientific, economic and political authorities, it is now the development society that speaks of protection of nature, ecology and new energy sources. But can there be such a thing as the protection of nature by the society that destroys it? To put the question is to answer it. By recycling ecology's slogans in this way, it is protecting itself from the reactions of nature and people. For, to continue exploiting the earth, it will increasingly have to conserve it, and our society will also have to provide placebos for natural human needs that are becoming exacerbated. Our society knows how to 'take action', feed the masses words and spectacles, and get the most demanding individuals jobs in labs, offices or parks where they will satisfy the quirks not permitted elsewhere.

Hence what can only be called the green lie: the mainstream discourse on 'quality of life'. There is only one way to give ourselves this luxury: by increasing production even more, so that quantity, swollen to the bursting point, gives birth to quality. One day, if chemical production increases tenfold, it will manufacture this miracle: rivers of water. But for the moment, we are still too poor. So, to obtain it in spite of the energy crisis, we must invest everything in nuclear power. Official ecology is also the alibi measure: the park or the nature reserve that gives over all the rest to bulldozers, the basin agency that protects water while multiplying irrigation reservoirs that deoxygenate or soil it. In the speeches of the INRA's new director, polyculture is gaining ground even as it is about to disappear in the countryside, and in the Paris offices of the ONF, hardwoods (which ones?) are reappearing while in the provinces the black tide of spruces is spreading; it is the machinery's inertia, people will say. In the verbal and scriptural production of official ecology belong the ministerial circular that circulates but does not impose anything, like the Chirac circular on the protection of seashores and great lakes, and the impact studies made after the impact, entrusted to the token ecologist, which have none on the building of a nuclear plant. There are exhibitions, like Protecna in Rouen, where polluters depollute themselves with synthetic chlorophyll and display their equipment for the purification and development of the seabed. Mister Renewable Energy organizes Solar Days where all the officials of the petroleum society may be found. But this kind of oratory central serves a psychological more than an energy-providing function, to reassure oneself more than the public. A ministry without financial or political credit caps it all. At least there are a few jobs in it for ecobureaucracy workers or the civil servants that the government sidelines in this garage.

However, official ecology cannot be reduced to this green smokescreen. For, beyond a certain point, development ends up devouring itself: once water is too polluted, it becomes unusable by polluting industries. But since they are designed too late, once the deed has been done, to unblock growth, protection measures only add up to taking one step back the better to jump forward. What is more, the conservation of nature is often hard to distinguish from its exploitation, for both are based on the knowledge of it. And it is realized according to the method and within the framework of the system. The translation of nature and of man into figures by science, thus making them digestible for computers, like that of free goods into francs and time into money by finance, turns them into interchangeable data or commodities, manipulable and tradable as such. The translation into concepts of the natural or human ineffable freezes it into a statistical or juridical element to be stockpiled and managed.

The recycling of ecology is thus achieved by the technocratic combination of capital and the State. Many profits (and thus jobs) can then be drawn from the protection (or rather the reproduction) of nature, that is, from the purification of air, water or landscapes with a lot of chemicals and electronics. A company like Sogreha in France derives important benefits from environmental studies, and the new university is also ready to work for the corporations. Environmental cleanup is now the third industry in the United States, and Lyonnaise des Eaux supplies France with costly purification material: it too cleans up what it pollutes, boosting its profits. With the oil crisis, we might think that solar energy is going to become a reality. But within the framework of development, it will not replace nuclear power but be added to it. And the day it becomes a genuine industry, it will no longer be the small do-it-yourselfers who will be in charge, but the State and the corporations that are already preparing for it. All this is eventually going to translate into a new step forward for the monopoly and its regulations, while the thoughtful ecologist does not know whether he should want them to be followed or not.

Thus, as a result of development itself, ecology seems to have the future on its side. In this increasingly specialized society, the scientific ecologist has his place as a specialist of the Whole, the controller of natural or human nuisances that might block the machine, the manager of water reserves, landscapes or peasants momentarily protected from developers. Industrialists, civil servants, academics, on your marks! The light is turning green, there are profits to be made and positions to be taken.

It is easy to understand how the serious *écolo*[7] would be tempted to leave an opposition that is futile for lack of means to join the powers that be, ready as they are to embrace him. And if he is a professor, let him dream of the millions that the State could grant his institute and of the positions open to his students. As for the tinkerer of solar water heaters, how could he not be tempted by Pechiney's billions? But the system has its own interests, as foreign to nature as to freedom. There is good reason to fear that the playing field is uneven between the ecologist and power. Will the first one be there to influence or to serve as alibi? It is not just any means he has been offered one drop at a time so far, and he will only be allowed to use them on the condition he does it within a capitalist or State framework that is not to be touched. Official ecology has limits which he will have to automatically stay clear of, such as development: if you are advocating solar, shut up about nuclear! Up to that point ... But no further. It is the rule that the ecologist active in oil, cement or TV practises instinctively. It is admissible to criticize sprawl when it is done by private citizens, already less when it is due to big developers, but never when State power is involved. You may decry the individual who builds a bungalow on the Crozon peninsula, but under no circumstances the submarine base that takes up a quarter of it. You may preside over

---

[7]From this point on, this friendly, folksy term is going to be used here rather than the ideological and spuriously scientific one of 'ecologist'.

an institution that buys coastal real estate threatened by secondary residences in order to hand them over to the Conservatoire du littoral,[8] but do not question Spatial Planning that smothers it in cement over 200 kilometres; be content to deplore such an error without being geographically specific. And if, as a single exception, your foundation buys a piece of land from private citizens on the flat shore of Hourtin Port where the MIACA would like to create a nautical base, do not say anything about all it is doing on the gorgeous craggy wooded shores across the water that the State is letting developers have. Flee like the plague from the screwballs who oppose it. Should you ever organize a colloquium on Aquitaine, do not breathe a word about its spatial planning. This word is taboo: it is a government matter, and you are only a State employee. Within this framework, one has all the means for acting. On the condition of not touching them.

If the energy crisis develops, shortages can sometimes paradoxically encourage development. We are short of oil? We must multiply drilling sites. Land is running out? Let us colonize the sea. The car no longer has a future? Let us bank on electronics, which will allow people to take imaginary trips. But we cannot indefinitely take one step back the better to jump forward. There will come a day when power has no choice but to practice ecology. Forecasting the future without illusions can lead one to think that barring catastrophe, the ecological shift will not be due to a tiny opposition without any means but to a ruling bourgeoisie, on the day when it can no longer do anything else. It will be the various culprits in the ruin of the earth who will organize the rescue of what little of it will be left, and who, after plenty, will manage shortage and survival. For these people have no prejudices, they do not believe in development any more than in ecology: they only believe in power, which is that of doing what cannot be done any other way.

If one weighs the costs of participation in official ecology in terms of both principles, one soon realizes that it is a fool's bargain. Either there is little one can do to change anything, or change is going to happen on its own when the time comes, as though man was but a pawn in the play of necessity and chance. Outside of any ideological bias, for simple reasons of facts and viewed in light of

---

[8]Translator's note: 'Coastal protection agency' founded in 1975.

its own ends, the ecological movement is condemned to break with the established disorderly order. It does not even have to seek to do it, it is going to happen on its own in reformist or revolutionary ways – it matters not – if he remains the least bit faithful to himself.

The critique and transformation of a society cannot come from itself; at most, its contradictions will help lead it there, if there be men to acknowledge them. The interested failure of the official protection of nature is due to the fact that it does not have outside itself a movement born of the depths of the people and of persons; to sail up such a stream, the rulers must at least be pushed with the help of a big stick. Even the reasons of science will be powerless if they are not moved by the powers of suffering and passion. Fishers, hunters deprived of fish and game, city dwellers stripped of space, air and peace, peasants of the earth, unite! If the protection of the earth is only a scientific, administrative or cultural matter that is discussed among polite specialists, it will remain the alibi of the society that destroys it, and those who accept playing this game will be its most valuable accomplices. One only defends that for which one is ready to get killed; and – at least for some – the human planet is worth it as much as religion or the nation-state. Nature is not a luxury, as some would have us believe; it is like water, air, freedom, which man cannot do without or else he dies. The best definition I know seems to me to have been given by an unknown foot soldier of ecology lost in one of those colloquia where technocrats and urban architects once again justified their job in the name of ecology: 'As far as I am concerned, if I can no longer fish, life is not worth living anymore.' Add to the river the tree and the house, and you will have the concluding word. Ecology is going to be saved only on the day when, leaving speeches, parks, colleges or colloquia, it will take to the street. It was born out in the open air, let it not be pent up in some office.

Recycling through spectacle

The recycling of ecological revolt can be done by the ways of cultural frivolity as much as by those of technocratic seriousness: they are complementary. In this case, instead of cooling ecological passion by channelling it into an administrative function, it is enough to *push it to the extreme*: its very outrageousness will make it futile, for, beyond a certain boiling point, it dissipates. Youthful passions will be allowed to flare up into the delusional. Decanted in some vacant lot, like literature or anarchistic journalism, they will evaporate without hazard and even be reusable for entertainment purposes. This inexpensive technique, similar to the lagooning of polluted waters, still entails a risk though: the accumulated mephitic gases can flare up, as May 1968 has shown, and the firemen then have to be brought in from Germany.[9]

This kind of leach field for the flow that could spoil the rest is culture, the realm of spectacle, where the nature and freedom shunted from life are backed up. The young situationists, who have since settled, had a deep insight when they denounced it but, unfortunately, they were content to feed it. On this page, this stage or this screen, there only remains a shadow – colourful of late – that lacks only relief. It is by now but the impalpable ghost of nature or freedom, only much bigger and more beautiful! It can be conjured up at will effortlessly; all that is needed is to push the button, take a look, or follow the guide. The beautiful waterfall or the seductive character whose role is entrusted to some star is now but the backdrop or the actor of a theatre where, having paid one's ticket – it is free for schoolchildren – one only needs to sit in one's seat in the stalls or on the plane. The earth becomes one big show business[10] where developed peoples – as in olden days Marie-Antoinette in her park of the Petit Trianon – amble along the alleys, casting a glance at a backdrop of cascades, rocks and ruins. And sometimes they are even made to play Indians like the queen played farmgirl. But to manufacture this fantasy for ghosts, virgin forests and real villages must first be razed.

The destruction of nature and freedom would be humanly unbearable if it was not compensated by a mirage that nothing hinders from being purely seductive, unlike daily life, which always feels somewhat rough to the touch. This is why economic technocracy is complemented by an organization of spectacle that,

---

[9]Translator's note: During the May 1968 crisis, President De Gaulle disappeared for a few days; he had gone to Germany to confer with the commanders of French troops stationed there, securing their support and making contingency plans for the containment of the revolt, so that for instance the elections scheduled for May 30 could go ahead without hindrance.
[10]Translator's note: In English in the original.

under the mask of fantasy, is also a heavy industry controlled by corporations or the State: TV is the best example. This function is the job of all kinds of stage managers: the stars of the various media, wandering entertainers or walkmen, band leaders or backup singers, park rangers or museum guides, cultural reanimators for dead countries or peoples, a public or tourists in a coma.

This theatre is the perfect spot for the passionate *écolo*, who will be able to use it to make a display of his intransigence and of himself: on certain specialized stages, he will be able to shout as much as he wants. Youth – and ecology has always had trouble getting beyond it – has always mixed sincerity with playacting, which it sometimes carries to the point of martyrdom: only age will make it possible to tell one from the other. Nascent revolt mixes with the deepest motivations the need to hoist one's person onto the stage. Individual solitude is never as great as during adolescence; what other way is there to assert it by sharing it? All youth makes man a seductive and seduced star who plays the role of his own person in front of his friends of both sexes; later on, one must be content with a stage shifter's position and to applaud, lost among the audience. Society, which needs stars for its televised theatre, encourages you to take a stab at being an actor. The more you will be extraordinary and uproarious, the better you will be received – provided of course you use this talent only at the theatre. And even on the stage, it would be best to avoid certain words that would fire up the audience and make it pour out into the street, as happened in Brussels in 1830. Up to that point (a small shudder goes through the room) ... but no further. This golden rule applies as much to the salaried ecologist of the Comédie-Française as to that of French technocracy.

CHAPTER TEN

Recycling through politicization–depoliticization

We live in a society that pushes to the extreme the division of tasks and zoning.[1] It finds in it a double advantage: the parcelling out of functions allows it to assemble these elements into a much more efficient rational whole, even as, isolated and mutilated, these gearwheels are no longer able to do anything by themselves nor to affect the whole. Science, industry, culture, in turn divided into subsectors, function in this way in their own realms for the benefit of an organism that then develops of its own accord. One might however think that one function manages the whole: politics, which is not yet a science and a technique, in spite of *political scientists*[2] who, before, during and after the war, have also themselves demonstrated how much they were in thrall to whatever climate of opinion happened to prevail at any given moment.[3]

---

[1] Translator's note: In English in the original.
[2] Translator's note: In English in the original.
[3] See, for instance, the works of Pr. Duverger from Munich to de-Stalinization.

From political commitment to withdrawal from politics

In principle, politics remains tied to the will of the people (otherwise known as *vulgum pecus*) and not to the judgment of specialists. But it increasingly becomes a specialized function managed by the party professionals who control the government.

In our industrial societies, politics in the original sense of the term, that is the governing – the orientation – of the social whole for conservative or revolutionary ends, is becoming well-nigh impossible. The social mass is too large and too heavy and, at the same time, too fragile due to being too complex. The instability due to perpetual change, the inertia of the technical and the economic make any long-term project inconceivable; governance takes place on a day-to-day basis, and in ways that are more or less smooth, we end up with the result we know all too well: invincible development, or underdevelopment if it happens to fail.

But if politics – the real politics of the mastery of the human environment and of its evolution – is becoming impossible, it is also due to another cause. At a time when man is no longer able to master his future, he is losing his reasons for having one, since the religious and moral engine that led him to govern his boat come hell or high water has lost all power. The death of man having followed that of God and values, where would he find the inner lever that allows Atlas to lift the world by doggedly flexing his muscles and his will? All that remains of it are the sudden flare-ups of a neurosis that fancies it can magically change everything at one stroke. Flare-ups that go out as quickly as they appear, and leave in their wake only despair and ashes, whose last embers will smoulder under the leaden screed of an inescapable everyday. If some day a crisis and propaganda stir them up, they will blaze up anew.

Although the achievement of social, spiritual and material ends through politics seems impossible, there still remains its residue, chemically pure for its part: power, now as the passion to wield it – or to submit to it, since it is the same one. Political commitment has always ambiguously mixed the desire to serve God, the public good or the people and the lust to climb up to the management of things and command over men: to exert political power which, more than ever, sums up all the others. Reinforced by this vice, great virtues are needed to sacrifice one's private life and take up a leader's responsibilities. Rare are those who have done it only for this motive, except for a few visionaries, at once caudillos and prophets, possessed by a rage for power that they call France or Allah. Too often, whoever devotes his person to the people devotes the people to his person as a sacrifice: one only has to think of a few recent examples. But for a long time, the obsession with power was contained within proper limits. To begin with, except for that of the priest, power was reserved for a caste or a dynasty and was only within the reach of an ambitious soldier during exceptional crises. Whereas in modern democracies or tyrannies anyone, in principle, can hope to reach the top. But then, since competition is unleashed and the road becomes longer and longer from the base to the summit, the conquest and conservation of power becomes the business of individuals and specialized organizations dedicated to this aim.

In a different way than in absolute monarchies, the throne is almost inaccessible in democracies. Any claimant must follow such particular detours, like elections and the rise within the party hierarchy, that attention to the means makes him forget the ends that were the rationale for his political commitment and identify any action with the struggle for access to power or the conservation thereof. The harshness of the competition between a host of candidates selects the most gifted for this specialized activity, the ones with a knack for planning and the manipulation of the masses. This form of intelligence can be the fruit of a narrow mind that does not question the value of stereotypes and the means necessary to succeed: if such and such a slogan works, he will be the first one to believe in it. In politics as in other areas, selection, being too specialized, takes place in reverse. Instinctively, the individual who has the least bit of intellectual or moral standards turns away from 'politics' by abandoning it, especially at the local level, to a collection of mediocrities devoid of talent for anything else. This is why intellectuals often have a contempt for politics that can easily go hand in hand with bad conscience about playing no role in it.

This is how a class of professionals of power is recruited, both right-wing and left-wing, but just as specialized as an engineer or a military man. By quantifying their results, the electoral media confirm these 'men of action' in the belief that they are dealing with concrete reality, just like technicians; and the mass of voters also fall victim to the magic of numbers. But these specialists of power for power's sake remain shuttered within their closed world, unable (and not having the time anyway) to cast a glance at the outside and any slightly distant future. Since they do not make but always follow the event, these realists are totally unrealistic. The change that is upsetting the lives of their fellow citizens escapes them: they have not provoked it; they ratify it and so precipitate it.

In industrial society, as in any other, the political function is organized. As the government is chained to its administration, so the politician is chained to a party whose principle, like that of any organization, has nothing to do with freedom and equality. The necessity of taking care of electoral campaigns and propaganda, the voting discipline in parliament freeze the tendencies that led voters and representatives to spontaneously gather into administratively organized, hence centralized parties. Electoral democracy, in which de jure everything flows from the basis to the top thus functions with de facto institutions whose principle is the exact opposite. As in the army, a leader and his lieutenants, recruited by co-optation or limited suffrage, command disciplined troops whose military are termed 'militants'. And like the army, the party has only one goal: victory over the enemy. To reach it, all means are valid; if need be we will do the same thing he does, only better.

The parties, their militants and their general have but one goal: their means. Since for the party nothing exists but the party, the one that is worthy of its name tends to become totalitarian and, to justify its absolute power with the people, it invokes an ideology as formerly the monarchy did religion. Ideology is only the party's means, not its end and, as in war, all conversions are welcome if they bring victory.

Thus, under forms than can be more or less bloody, the same enterprise continues: the development of political, economic and technical power, with all its side effects. This is why, since we are stuck with it, the liberal way of America or of social democracy is preferable to the totalitarian way, whether red or black.

Except when it comes to manufacturing power, the political machine runs on empty. This truth hidden until May 1968 is starting to come to light, so that depoliticization is now succeeding the politicization that prevailed until then. The war, of which revolution is but a chapter, had brought about in France a mobilization of minds that was all the greater as it had happened belatedly, after the Algiers landing and the Liberation. The Right being busy with the economy, it became accepted among the left-wing intelligentsia that any action came down to political commitment and all politics to those of the Communist Party. But the Stalinist Church's sympathizers had trouble getting past its narthex. And since the USSR could not directly confront the United States in an atomic conflict, in the West this political commitment drove one away from power. With the Left shut away inside the Communist freezer, neocapitalism could manage development as it wanted. Besides, if social democracy had gained the majority, as in England for instance, it would have done the same.

Hence, after Stalin's death, Budapest and the belated revelation of the Gulag, the disillusionment over China, Cambodia and Vietnam, the depoliticization of the intelligentsia and of student youth, the usual vanguard of social change. The retreat to limited but concrete interests, the need to change life instead of a regime that promises an eternally brighter future and always ends up re-establishing the same order, only refurbished.

Then there is the ecological movement. Putting into question the entirety of the industrial world in the name of the preservation of nature and freedom down to the humblest details of daily life, it can bring back content and meaning to politics. But on the condition that it does not succumb to the mirage of power for power's sake.

Recycling through depoliticization

Industrial society secures its rule by concentrating all power in the centre: by politicizing. But it also manages it by depoliticizing, by scattering away at the social and territorial periphery whatever it cannot successfully assimilate. In other words, it exiles; but it can do so only if this exile is voluntary.

Politics, aside from the ups and downs of – oh so – foreign affairs and armament tied to economic and technical progress, is devoted to development (this goes without saying) in the name of raising the standard of living and then, as the latter becomes uncertain, in the name of employment. As for the impact on people's lives – no studies are available yet about that one – it is not the business of the various general staffs, but of the foot soldier who registers it in the trenches of everyday life: no more than about the war does anyone ask for his advice about the highway that is going to run through his village. As for the elected officials obsessed with traffic jams and employment, even today, the Infrastructure Ministry is forced to curb their boundless appetite for parking lots. Thus, the people affected began to realize that the fate of their living environment (*not* their living standard) had little to do with the play being performed way up there at the centre and on which they had only an illusion of control every three or four years, while their destiny is at stake and being decided on this very spot. By defending their tree or their house, citizens are learning again this vital minimum of civic community that consists in getting together to defend one's interests. In defence committees (*Bürgerinitiativen* in German), bypassing the political representatives always missing in action when these issues come up, they directly take charge of their environment by using the law of 1901 on associations, along with the other legal guarantees from the Third Republic that the Fifth one has not abolished in the name of technocratic efficiency.[4]

But defence committees have the vices of their virtues. They do not defend universal and abstract entities but only immediate and concrete – hence limited – interests. They thus find themselves being countered with the General Interest, and often end up losing when the Superior Interests of France (SIF), be they infrastructural, atomic or military, are at stake. Not seeing any further than the horizon of their parish, they are down-to-earth, particularistic, and wary of strangers. 'They don't do politics', they are afraid of calling into question the government and the regime, and it is only reluctantly that they get angry with the prefect. Still, these committees are a real nuisance for developers and the administration, as the latter sometimes invokes them to curb the former. If they do not always win, they score points, delay or limit the damage; above all, they break the silence: where it reigns, one can be sure of losing. Defence committees are getting Frenchmen to learn anew what is genuine politics: the one that preserves or changes daily life. They teach them to associate and to take charge of themselves in matters that

---

[4] This is why a decree published on 29 July 1979 in the Official Journal modifies legislation termed 'outmoded'. 'With the just-published text, the State is now in a position to respond with appropriate legal means to legal action opened for instance in connection with the building of nuclear plants.' *Le Monde*, 30 July 1979.

directly affect them. And, proceeding from their particular case, to grapple with the real issues of modern society. Local action can lead to overall views, while the obsolete ideologies of parties are only blinkers.

Depoliticization also leads one to immediately change one's life in a sect or a community on the margins of society as a whole, which tolerates and even encourages them. Since it cannot digest this foreign body, it encysts it, playing on its very hardness that makes it unable to spread through the whole. The counter-society is thus a victim of its will to resist its enemy's grip: it turns, as it were, against itself and inwards the energy it cannot spread outwards; without knowing it, it collaborates in this way with the order that forbids it to change anything to this society as a whole. This process that turns the counter-society into a sect or a marginal community can be termed sectorialization: the sect being set apart as much by its nature as by its place. It may be religious (see Jehovah's Witnesses, the Friends of Man, Jones' Peoples Temple, etc.). But also political–religious (see Trotskyite or anarchist groupuscules), or even political–literary like surrealism and its double, situationism, which has had a part in the ecological movement alongside naturist sects.

The sect compensates its limited power by the intensity of its authority, the possession of truth and the certainty of embodying it by its virtues, hence its puritanism. Like mass parties, it takes pride in its small number of elect (they are chosen by the Lord or by History rather than by the electorate) and in the rigour of its doctrine, whose logic it pushes to the end with no concern for experience. This is why the sect often appeals to intellectuals as language specialists. Invincible, it defies time and has nothing but contempt for current affairs that invariably confirm the one Explanation. In our time of squalls, the sect thus provides individuals lost in the storm with a warm refuge where, huddled together, it is possible to feel sheltered. Society as a whole encourages it to thus invent for itself a secret kingdom where its excesses will be harmless. Ideology would be too dry if it was not vivified by mysticism or the spectacle which surrealist and situationist sects have indulged in.

As much as by opportunism, the ecological movement runs the risk of uniting and dividing itself into chapels or sects of the pure who will cash their salaries in the guise of a righteous conscience instead of money or honours. The temptation is great, especially for a minority of young intellectuals, to oppose totalitarianism and industrial fundamentalism with the radical 'no' of a fundamental naturism. This is understandable: since current society now only leaves a choice between itself, tending to become everything, and nothing, the opponent chooses this nothing, which is convenient for all involved. So everything is urbanized? I refuse the city. Everything is schooled? I refuse schools. Everything is State-controlled? I refuse the State. Everything is conceptualized? I negate language and so on. I am off. There is a risk of getting 'fed-up', which one must be wary of. We answer the implicit dogmatism of the industrial world with explicit dogmatism: the compensation for powerlessness is the ideology that wields in heaven the power that Caesar monopolizes on earth while waiting for the revolution that will make it possible to combine the two. And since life imposes compromises, the pure find at every moment other pure who exclude them. Since the possession of the true doctrine leads to the condemnation of heretics, sectarianism drives apart into chapels, each with its own pope.

The sect counters mainstream stereotypes with its own, following suit with its own taboos, authorities, powers and internal disciplines. One who, refusing the order of the whole, enters this kind of society, finds another order, only more rigid. Ultimately, wherever such a thing is tolerated, the sect constitutes a totalitarian microcosm in which are unleashed, along with power, idolatry, money and, as the price of final ecstasy, mass murder or collective suicide. All this makes society as it is the only guarantor of reason, freedom and respect for man. But the worst of all regimes is the one where the sect becomes the State. That day, we may be sure that, since absolute truth excuses all means, the violence inflicted on men and nature will no longer know any bounds.

Fortunately (or unfortunately), the current system hardly leaves any place for social creation, apart from its craziest forms, as if it knew they were useful to it as a foil. The first criticism that can be levelled at the community founders, apart from their illusions about rural life, is that the French countryside is not the Far West of the era of Phalansteries: this space that has been desertified by technocrats no longer has anything to do with the free desert of pioneers; it is invisibly canvassed by the power of administrations and of the SAFER.[5] And the tithe that the State collects, bringing

---

[5]Translator's note: Acronym for *Sociétés d'aménagement foncier et d'établissement rural* ('Property development and rural settlement societies').

back to itself any social function, is such that it reduces to nothing the surplus that would make it possible to build a polity beyond private life. Pretty soon, to do a revolution, we are going to have to apply for subsidies.

Understandably, society as a whole has no trouble putting up with this kind of fixation abscess. And many ecologists get tempted to occupy this pathetic 'niche' that is offered to them. Society feels less threatened by small groups of the pure than by a more general, more flexible attack. We have seen that, in various guises, the ecological movement can be assured of having its small chapels in the industrial cathedral; and if shortages are there to stay, ecologists will be entitled to a stool or even an armchair by the throne. This leaves but one way to avoid getting recycled: instead of fleeing to the margins, striking at the centre.

Recycling through politicization

Politicization is followed by depoliticization. But the latter soon reveals its limits. The defence of nature soon raises all other problems, be they economic, social or, ultimately, political. One comes up against entrenched interests and the powers that be: capitalism, government, the army or parties. The very success of the ecological movement gives it a political dimension. Some of its members dimly feel the need to get it out of its ghetto. Afraid of seeing it mystified and neutralized by the system, they talk of politicizing it; and those *écolos* who have gone over to politics and those party and union militants who have gone over to ecology agree in complementing it with a socialism. Others, conscious of its distinct calling, are thinking of founding an ecological (or ecologist) party. But in getting out of its ghetto in this way, is ecology not running the risk of losing whatever made it different?

The ecological movement is getting old. In its adolescence, it could believe in changing life immediately, but it is now discovering power, and some go so far as to take up the 'everything is political' of certain parties. This formula is inspired by the natural need to translate truth into acts and by the sense of a glaring emergency, but also by a society that opposes thought to action, defined once and for all by adherence to such and such a party. We must act ... And acting is 'doing politics' (your enemy, who agrees, will tell you that a free spirit 'does not get involved in politics'). And doing politics is at most defining a program, all the more serious as it will take the form of a few brief articles, the result of an expertly balanced compromise reflecting power relations inside the party. Even then, this tends to be a waste of time. Acting mostly means spreading the gospel among the masses as circumstances dictate: which quickly becomes propaganda that one is the first to fall for. It means preparing elections and, to this end, finding funds and setting up an efficient organization, hence of necessity a rather centralized one: in short, forming a party if one does not want to join the one that has proven itself.

For politics is reality; everything else is the idle chatter of intellectuals. It is thus the one that exists rather than the one that one invents. This leads one to introduce ecology into the big parties that have the means or, better yet, into the government, where the possibilities are even greater. Ecology is then no more than a card in the game of Right and Left, of reformism and revolution, and it will be welcomed everywhere with interest – and, above all, with the hope of inheriting the 4 per cent of votes that give the majority. The easy-going ecologist, with little taste for ideology, will be tempted to collaborate with the centre-right in power, but usually he does it 'without doing any politics' within an official or private institution. If he is left-wing, he will join the Socialist Party, close to power albeit always kept out of it, and whose vague declarations allow everyone to read their own truths into them. But when it comes to the government or to the big parties that can aspire to it, one must have the ability to be content with a few measures or, more likely, some dramatic, fine-sounding words. This is selling very cheaply the Green label the public now demands. There is greater hope at the PSU and the CFDT, which are more open to the issue because they are further away from the responsibilities of power. But even then, ecology is a belated afterthought; would it not be better for it to turn into a political party itself?

A genuine ecological politics assumes, at the same time as a reversal and a synthesis of the values of Right and Left, a change in political and social orientation that parties have been incapable of achieving. It is not a mere strand within the UDR,[6] the Socialist

---

[6]Translator's note: Union des Démocrates pour la République, a Gaullist party that existed from 1968 to 1976.

Party or the Communist Party. An ecological politics is something very different from a dab of green on pink or red, or even a stroke of white or red on green, that is liberalism or socialism, slapped onto ecology. The latter has no business in the family quarrel between industrial society's right and left wings. Hence the idea of the ecological party, which however has asserted itself clearly only after the 1979 European elections. But can organization into a party and ecology be compatible? If every precaution is taken to make the latter as little centralized, hierarchical and disciplined as possible, as has been the case until now in the ecological movement, is it not going to be made inefficient? And if it is strictly organized, will it still be ecological? Are we not going to fall back into the rut of the old parties? Before it is launched, reflecting on those issues would not be a waste of time.

Heaven knows that the 'everything is political' has proven itself over two centuries of wars and revolutions. It is reminiscent of EDF's 'everything electrical' – where, too, it amounts to putting all one's eggs in the same bottomless basket. The 'everything is political' soon gets impoverished into 'everything is electoral'. Or worse, in case of violent conflict, into 'everything is military'. When it comes to elections, the most important of which are the national ones, it becomes 'everything is national'. The political vantage point, in the current understanding, ends up identifying action with the conquest of the nation state, that is, with the type of society that the ecological movement claims to be fighting against. To the extent that one seizes power, power seizes one; once one is in power, one remains there. There is no example of a State withering away by itself after a revolution; after a while it rots, which is not the same thing.

Power deceives. Political action dispenses those who practice it in government or in the opposition from thinking and acting, this time for real, that is, not to adapt to circumstances but to master them. How many people think they are helping others when in fact they are fleeing private issues and a more immediate reality that concern them! Thus, 'politics' accelerates current change without changing this change in any way, and for over a century, not only has it looked away but it has also turned away public opinion from all the change brought about by development in the environment and in life.

To be sure, involvement in the electoral process has had some advantages for the ecological movement. It has brought it publicity and given it access to the media. But not without having to give away something in return. 'Politics' is a demanding deity. It gobbles up a fledgling movement's meagre resources in cash and militants. Be it only to print professions of faith, the funds have to be found, and one is tempted to ask corporations or the State to contribute, so that, implicitly or explicitly, one becomes dependent on them. The time and money devoted to electoral action may well be missing for other activities, such as the work of keeping the movement and the public informed. If one is not careful, the preoccupation with politics and elections becomes a kind of cancer that absorbs everything. The electoral result is the judge of last appeal: another 1.5 per cent and life finds its flavour and its meaning again.

Political commitment gobbles up personal life, without which it loses all quality, since the quality of an act is but the fruit of a person's reflection and virtues. And it also risks developing at the expense of the creation of new social forms in communities and in local defence actions. Finally, political extraversion takes one away from all retreat and reflection, indispensable as they are if one expects a movement to remain faithful to its ends because it is capable of judging its means: without constantly going back to the wellsprings, action, in a different way than discourse, is but empty noise. It is not in discussions within commissions and congresses or in electoral meetings that the ecological movement is going to be able to put together a true program for, be it internally or externally, power is at stake there. The debate is skewed because there, too, time is running out and, instead of discussing, people throw arguments at each other's faces like stones, or by way of answering one's interlocutor, one resorts to slogans hurled above his head to feed the public's prejudices.

The pursuit of power leads to the constitution of organizations, which are all mass organizations to one degree or another – how would it be possible to do otherwise if one does not want to get lost in the mass? – hence are willy-nilly ruled by a leader which the led dream of toppling. There is no such thing as a democratic party; the best one can hope for, as for the State, is that several strands (or gangs) fight over its leadership. This kind of institution runs the risk of putting at the movement's head a race of activists – one could also say of high flyers or social climbers – gifted organizers of things and manipulators of men who will soon eliminate the true ecologists. And the struggle for power will rage within the party between strands among which ideology hides personal rivalries. The short history of ecology has sufficiently shown how much political commitment fosters these kinds of laughable, paralyzing quarrels. Power corrupts ... Power maddens ... No need to be Stalin or Hitler, they circulate by the thousands in more modest editions. Should the ecological movement forget it by becoming no more than a party that might take power one day, we may be sure that, once again, the obsession with holding on to it will make it continue in the same rut, in a harsher way, under a green flag.

Political and electoral action is legitimate, ceasing to be so only when it becomes totalitarian by claiming to sum up all action. Politicization then becomes a kind of disease, in the same way as depoliticization, to which it leads back sooner or later. Refusing the dialectics that swing from communitarian disillusionment to political disillusionment and vice versa, the ecological movement must act in all areas, starting with those within its immediate reach. The *écolo* who has a political calling need not worry; if he pursues his reflection and his work, very soon – and it is already the case – he will be forced to deal with political issues. For if social change implies political change, it goes much further. The ecological revolution is not meant to change a regime or even life, but a world.

PART FOUR
Fruits: Sketch of an ecological politics

Now comes the step of action, which is the fruit – hence the seed – of the spirit. But, *pace* the late Teilhard, any human phenomenon is a leap, a break in evolution. There is thus nothing necessary about it. Anything can be won or lost. Man has nothing to rely on but his own freedom from now on. It is up to him to choose to live on earth.

CHAPTER ELEVEN

Topical utopia

Now what do we do? The reader who has followed to the end the thread of the demonstration should already be able to work this out. Contrary to the preconceived idea prevalent in a society that wreaks havoc with the planet in the name of 'positive' action, by which it means a purely economic and material one, any question that is properly formulated already opens the door to its own solution. The analytical negative determines the practical positive. Critique is the mould; it hollows out in the hardest metal, which is also the most sensitive one, the relief that will later follow its outlines. So now it ought to be for the reader to draw the conclusion. But while we're at it, both of us might as well get down to it.

Ecological conversion

To some extent, the ecological movement tends to constitute what used to be called a counter-society in May 1968. Each of these two words is as important as the other.

Counter. Any revolt, and especially this one, is a 'no' shouted at the social state, and this one is far more radical than any other because this time it is not hurled at such and such sovereign or church, but against a world that tends to coalesce into a scientific, technical, economic system that ends up being statist, bureaucratic, military and police-run, where nothing escapes Leviathan's eye and hand. A cultural totalitarianism of which political totalitarianism is but the more or less necessary conclusion. If the latter does not yet control the totality of earthly space-time, the former is in the process of imposing a single type of culture, lifestyle and thought to the near totality of the planet.

Society. Hence the response, that should not turn into a duplicate, of a cultural revolution that would not only change the constitution, but also the social whole in all the concrete details of its existence: its culture in the ethnologists' sense. Against a kind of social avalanche that threatens to destroy, if not control, the totality of the human world, or in other words, nature and freedom, the ecological movement finds itself compelled to counter it at every level of social life with a radically different project. It too is not allowed to forget anything – neither the economy nor politics, neither contemplation nor manufacture, neither leisure nor labour, neither the sea nor the kitchen ... An endeavour without precedent, like our times. Since the social state is for the first time taking a totalitarian cast, whoever refuses it is compelled to reject society, unlike the revolutions of 1789 and 1917 that hardly put in question the foundations of morality, the economy and the State. For them, it was a matter of doing things better, while for the *écolos*, the issue is above all to make a clean break.

And yet, the ecological project has only one goal: foiling a totalitarian enterprise that is all the more dangerous for encompassing the entire planet; unlike that of Hitler and Stalin, it does not even have to say it. The ecological movement is compelled to envision a total revolution for the exact opposite reason than the one at the origins of political–religious totalitarianisms: out of a refusal of absolute power, both spiritual and temporal. Its overall project is not – and must not be – dictated by the passion to know and control everything; although inspired by the love of nature and freedom, it is determined by the situation and by its adversary. Ecological revolution, like our era itself, has nothing to do with those that preceded it. An action against the decomposition or the freezing into a single block of the human planet must therefore never lose sight of its own paradoxical character: that of being total, but not totalitarian. Which leads one to assert many other contradictions or paradoxes.

First, the 'against'. This term, shunned by every society, and especially by the one for which everything is war, is however generated by a 'for'. It is out of love for the basic reality, life on earth, and the loftiest human dream, to be free and to become free, that the ecological revolt has been driven to radical conversion – that is, an about turn. Ecological action is the very type of revolutionary action, if this means going against the grain of society's state and becoming. This is why it has no reason to take on the air and the phraseology of modern revolutions (?). It is already disruptive enough by its very nature, alas.

Simply by demanding respect for the earth's ecosystem that has allowed life and man to appear, ecology challenges the social state far more than socialism and communism as heirs to the values of industrial and bourgeois society, content as they are to claim for themselves the control of the State and the socialization of economic development. It is not only the religion of profit that it rejects, but also that of production and cost effectiveness, not just the reign of multinationals, but also that of industry. The principle of ecology is subversive even if the ecologist is moderate, which he can remain only by cheating with its truth. One starts out defending trees and cute animals, and then ends up coming up against the CEO and the prefect. One fights for a marsh and against a housing project, and then ends up questioning population growth. How would it be possible to drop such and such a project for a marina or a factory without also dropping the taboo of employment? Pollution is denounced, but how can an end be put to that of the Rhine without also raising the issue of Europe? And that of the ocean without pondering the World State that alone seems able to prevent planetary catastrophe? If we want to save nature, how can we avoid entrusting its management to a synarchy of technocrats like that of the Club of Rome, which would be its exact antithesis? How can we imagine an international institution having any real power that would respect the freedom and the diversity of peoples and individuals? Starting from the experience of local facts, one ends up with universal issues of the human condition. Ecology is relentless; it leads you straight to ultimate questions about the meaning of life and the social contract.

The core issues of ecology are indivisible; it is compelled to go all the way to the bottom of the issues raised. It cannot put into question the earth's devastation by man without tackling its *active* principle, science, and its heir, technique. The explosion of human power is explained in part by the fact that the operational truths of science have taken over from the mythological truths of religion. Human energy, a good share of which used to be dissipated into heaven, has been entirely invested into the earth. And these rational and pragmatic truths, at a time when the ideological or religious particularisms threatened by technoscientific unification are heightened, tend to become the only ones that can be acknowledged by all men, and as such, the sole basis for internal and international consensus. This is why, in the current state of things, aside from the war of each against all, the only order conceivable is that of a total world government of the earth in terms of scientific knowledge that would not forget any factor, including the human one. This scientistic utopia is not a view of the mind; for anyone who accepts the currently given, it is that or nothing: chaos, atomic death ever-better equipped by science and technique. If human freedom fails to deal us a new hand, we have but one choice: scientific war or peace.

The ecological movement is therefore condemned to raise the issue of modern freedom. Where there are no longer any religious truths, how do we prevent the operational truths of science, and of its extension in technique, from taking on an absolute value? Which leads either to the destruction of an overexploited earth, or to that of freedom by oecumenical management in the name of scientific authority. The ecological movement is not going to bring any change to the current evolution if it fails to relativize the truths of science, by *legitimizing* its sphere without negating it: by distinguishing its share, by establishing the sacred boundary markers – both positive and negative – that it cannot go beyond without annihilating man along with nature. This task is particularly pressing when it comes to the human sciences, as they claim to take man as their specific object. At least insofar as they deserve their name and are not simply the ideological or mythological reflection of the age.

The authority of science can be curtailed only in the name of a higher authority, be it intellectual, moral or religious, it matters little; the main thing is to know which one, and that it be acknowledged. The author would be presumptuous if he got out of his pocket this truth that would allow man to make his way between the sacred totalitarianism of the past and the profane totalitarianism to come. He is but a man, entrusted for a day, despite himself, with a spark of life that makes him talk of nature and of freedom. If he thought he was the only one, he would remain silent, but others before him have sensed them, such as those living atoms – all the more radioactive – who have constituted the ecological movement. These two words are only the signs at the start of a secret path. But they open it to the one who stops, silencing any other voice in him.

If the supreme authority of our time, science, was not problematic for the ecological movement, it would change nothing. It therefore cannot elude taking a distance from science in the name of its principles; this is the first condition for a mastery of technique without which a new orientation of economic and social reality would be an idle word. It must do it all the more as this distancing leads to a liberation of science itself, since the objective – and subjective – awareness of the mechanisms of universal material power forces science to be married to the corporations' and the State's economic and political power, full stop. The current threat in all its forms, including the atom bomb, is the fruit of the knocking up of this feminine Knowledge by masculine Power. Thus, doing the opposite of what has been done so far, an ecological politics should disconnect science from industry and the State. Which leads to the abolition of the management of scientific research by giving back to science a freedom that is that of humility and poverty. Technoscientific development will then stop being so explosive (in every sense of the word). Its pace will be slower, which will allow us to know and master its effects: the impact study will no longer rubber-stamp the fait accompli. Instead of indefinitely adapting to a change that overwhelms him, man will be able to make projects that more powerful means enable: no convivial society without soft technology, but no soft technology without soft science. However, active tempers uninclined to philosophical reflection need not worry; in the case of ecology, the break on principle is inseparable from a break in actuality.

No more than nature is freedom an ideal that encourages us to flee the earth, unlike socialism, which, driven at first by a passion for justice to change the proletariat's present existence, degenerated into an ideology that promised the Absolute for the day after tomorrow. On the contrary, up to now, the ecological movement remains close to its source: the humble joys of earth and the aggressions it undergoes *hic et nunc*. Its action is inscribed in space-time or, in other words, in reality. It does not oppose only a juridical principle that would sum up all ills like property, but along with the cause of all these effects, opposes their very multitude: this military camp, this nuclear plant, this dam. While it promises annihilation to the bourgeoisie for tomorrow, the Communist Party leaves it everything for today – which is all it is asking. Whereas, along with development, what the ecological movement challenges is the only principle that truly matters to neocapitalism: its wallet. It does not talk about profit, it does better: it attacks profits. It can afford the luxury of moderation, as it cannot elude an unremitting conflict. For if there is today a power that does not forgive, it is financial interests when they are threatened.

A mediation between opposites

From the changing surface of the sea to the dive to its depths, ecology's approach is thus the reverse of that of a religion or an ideology that starts from its truth to explain everything. As with other goods, knowledge of nature and freedom is given to us by the infliction of their deprival. The ecological experience has little to do with the ecstatic revelations of religious or political mysticisms, for it is essentially earthly: at once raw and reasonable. Though it arises out of depths of which it initially knows little, the power that animates it makes it hit hard the obstacle that blocks its path, and its first feeling is one of suffering and anguish. It discovers itself to be in a contradictory position, that is, alive. The question that the ecological movement raises and asks itself is a fundamental one, encompassing the human universe; it may thus be termed a religious one. However, it has nothing in common with more or less declared religions which, by bringing the final explanation and solution, forever exempt man from seeking his own truth. It is not for ecology to provide the mystical or ideological gadget that would reveal and magically reverse the course of things; this would mean betraying its calling. It would be regressing to childhood, sterilizing its reflection and dividing into sects. Its raison d'être is not to preach the alpha and the omega but, from one to the other, to firmly hold its path. Between hell and heaven, its domain is the in-between: the oecumene in this fin de siècle. Nature and freedom motivate it, but if we ask it why it can only say: that is how it is, without air or water, or the opportunity to be even remotely myself, I will die. To be sure, starting from there, one can add a host of rational arguments, some of which have been presented. But these are but add-ons generated by that vital force that makes you cry 'no' to the world that is besieging you.

The function of the ecological movement is not to teach the Truth, but to remind us of a crucial situation that is revealed today by any actual passion for nature and freedom. Anyone who feels it knows that this principle also holds for every man or for none. In other words, it is about the right for any species, individual or society to live *its own* life: its difference. Hell is but a single Nothingness. The ecological movement thus has not invented anything, having been content to discover there as well a truth that is on the market as much as on the savannah. There is nothing more banal, and yet more paradoxical: the common good that unites us is what distinguishes us. It is because an oak tree is not a beech tree, because Paul is not John, because Quebec is not Euzkadi[1] or France, that their cause is the same. A planetary cause, that of the sole planet that is not plunged in mourning by the same old shroud of ashes or lava, but which, at every step, makes a new flower blossom. There can thus be no question of turning Biafra or Languedoc into a nation state, complete with its own steel mills, colonels, statistics and H bomb. We are not united by a flag or by a little red or green book, but as allies can be in a just war, if there ever was one. No matter what is our church or our skin colour, Jew, Muslim, Frenchman, Gascon, Christian, atheist, liberal bourgeois or socialist proletarian, there will be no more room for us on a planet deprived of air and water than in computer-managed survival.

At its bottom as in its details, the ecological struggle is the antithesis of a crusade. It is not joined to establish the Heavenly Jerusalem, but first of all to defend what is, this time in the widest, strongest sense of the word: nature and culture, one and manifold. Besides, before attacking, one must defend oneself. Ecologists would probably disagree about the ideal society, whereas they spontaneously agree against the common enemy. On principle and in effect, ecological union can be neither a system nor a church but an alliance. Originally, it does not gather people united in advance because they belong to the same group, but brings together a diversity of individuals associated precisely to defend the right of each to lead a personal and social life that is different. But this factual association, this wartime alliance is aimed this time at that peace which, until now, has marked the end of coalitions. A paradoxical enterprise which consists in uniting men and their societies by the very reason that used to separate them.

---

[1]Translator's note: The Basque name of the Basque Country straddling Spain and France (on which side Charbonneau had his Pyrenean winter house).

Ecology does not bring *the* Solution, but posits in each case the contradiction that freedom must resolve for a time. It is not immobility, but perpetual motion: progress; we may now take up this word again for our own. From the start, it must bear the contradiction between nature and freedom that is its life and its wealth, and it forces it to find itself everywhere and every moment in conflict with the industrial world. At every step, ecology finds itself caught between opposites: what it wants and what is, an emergency and the long term, society and the individual, State and society, reformism and revolution, Right and Left ... And what not? One would need to circle the earth. Halfway between the sacred Truth and Order of yesterday's religious societies and the logical and pragmatic truth of tomorrow's scientific order, ecology's time is that of freedom.

While religions and ideologies polemically oppose the terms of the various contradictions so as to abolish them by the magic of discourse followed by that of arms, ecology must endure them regardless of the anxiety they give it, because it is up to it and to man to effect the mediation. Not the compromise, but the decision inspired by the spirit, which depends on the case and the circumstances and is always difficult because it sacrifices some possibilities. To act is to choose: thus, nature to save it from man, or man to save him from nature. Ecology does not have any miracle solutions to offer, but choices which, even when they have been carefully thought out, still cut to the quick.

A revolution for that which exists

The ecological movement's visibly very undoctoral doctrine could be called a topical utopia. Born of a desire without end, of the dream of Eden, of perfect freedom within a nature reconciled with man, it is also the product of a historical situation: the refusal of a social supernature that tends to destroy the one and the other. U-topian in its deep motivation, it is also inseparable from its place and time. It is 'committed' to the reality of a unique present, but in an altogether different fashion than those governments or parties that are stuck in current affairs because their mind is still moving in the nineteenth century. The ecological enterprise is essentially realistic; the utopian one is really the technocrat or the politician who remains blind to the effects of his achievements and insists on wanting to stick infinite development within the finite.

While it recruits its militants in a leftist milieu that dreams of a perfect society, on the ground the ecological movement defends what is: those trees, those villages, this culture. It is at once revolutionary, because it demands of society a radical change of direction, and conservative: at every moment, we find it in the paradox. It must not be ashamed of being conservative, far from it; it must wrest this term from a Right that no longer conserves anything of the treasure accumulated by the earth and by men. To hold this front, one must move differently, spiritually and physically, than hurtling down with this avalanche called 'order and progress' – 'disorder and regression' would be more accurate. To defend what exists against the chaos of the system, the *écolo* will never have too much freedom, nor too much willpower to break the moorings, nor too much imagination to tie them.

The ecological u-topia has but one locus where it may become something other than a lie: the earthly *topos*. Why do we have to remind ecologists of this? It therefore has no greater enemy than the idealism that verbally identifies the ideal and the real, the intimate enemy whose temptation lurks at every moment. For it can only lead the ecological revolt to failure, crushed or hijacked. It has but one way of escaping the lures of technocratic efficiency or leftist show business: a genuine realism that judges – but does not justify – things as they are. Where else is one going to start from to change them? This is what must be called seriousness, which is not to say that humour is uncalled for when matters become too serious.

As a child of the May feast, leftist ecology persists in taking its desires for reality. Twelve years later, it remains an adolescent, with the weaknesses of this age that demands a lot, out of an excess of force and because it has not yet been confronted with the experience and responsibilities that crush its elders. What is a normal attitude at twenty is no longer so at a later age. To artificially prolong youth is to betray it by denying nature. Ecology teaches that a fruit must mature; so must the ecological movement if it does not want to wither or rot.

'Take your desires for reality' is not inspired by the spirit of youth, but by an old devil who wants to make it abort. It is a sterile slogan,[2] best left to the 'men of action' who do not see beyond the

---

[2] Translator's note: One of the most famous graffiti to sprout on Paris walls in May 1968.

end of their noses, like Mr Boiteux when it comes to nuclear energy.[3] Making a reality out of one's desire is something else altogether. This assumes that one is not confused with the other in the first place, and that one ask oneself into which reality a given desire – not just any desire – is to be translated. To succeed in this, there will never be too much of the old man's caution, united to youth's *élan*, to go around.

Otherwise, the ecological movement will seamlessly transition from the childhood disease of idealism to the senile disease of narrow realism. Which is what happened to many other children of May. Some have been destroyed by their desire, but most are now family heads, among the unemployed or the workforce, or members of the establishment. The latter will be even harsher bourgeois than their fathers. Because they chose the ideal identified with reality, they have chosen reality (reduced to career) identified with the ideal. Whoever wants to escape this eternal return that leads from an extreme-left infantilism to a senility of the extreme centre has but one way of avoiding it: not to identify what should be with what is. It is the only way to bring the former into the latter. In human terms, maturing means to become an adult. Not to repudiate, but to accomplish one's youth; there is no other way to be faithful to it. This is to confront life, which is contradiction. If the ecological movement does not shrink before its own contradiction, namely, the tension between freedom and the universe, instead of just keeping to either one of these terms, that contradiction is going to be resolved both in theory and in practice.

It must therefore recover from its childhood disease: extremism; driven to a decisive 'no' on substance, it has no reason to become intransigent about forms. Struggling against an extreme peril, it has no need to add to it by cultivating spectacular extremism. On the contrary, a relaxed and calm attitude, an openness to people and to facts are necessary to it. It has too many 'no's to say to yell them. The ecological movement will be able to make any

---

[3] Translator's note: As head of Électricité de France from 1967 to 1987, mathematical economist Marcel Boiteux was a key architect of the development of France's nuclear industry. A bitter irony for Charbonneau was that Boiteux had taken part in his personalist circle in Bordeaux in the 1930s, and had thus been exposed to his early critique of technique and industrial development, to little avail as it turned out.

progress only by eliminating the element of neurosis that the loss of nature and freedom feeds within it. And that which any radical action undertaken against the state of things attracts and solidifies. Neurosis is selfish; deep down, ecology does not interest it, it is only for it an opportunity like any other to act up and exhale a fury whose motivations are much more psychological and private than political. Neurosis is conformist; it never threatens the society that allows it to shout or even enjoys this spectacle; it can only deceive and paralyse a movement, lead it astray in needless violence. Such a fundamental social change is not a happening, nor even a festival meant to purge society of the drives that threaten it; it is a huge undertaking over the long haul that requires at least as much objectivity and reason as the management of the Régie Renault.[4] Therefore, in reflection as in action, we absolutely must get rid of this kind of dross.

The ecological enterprise can be termed a conservative revolution. (If you are of the opposite persuasion, call it 'revolutionary conservatism'.) It is not subject to the traditional opposition between movement and resistance, progress and reaction, reformism and revolution. The best of its freedom is in its experience of that which exists: the natural environment and daily life lost sight of by the scientists, ideologists or politicians of industrial society. That which exists is the turf in which the ecological movement has sprouted and must continue to plunge its roots if it does not want to wither. Its defence committees are as necessary to it as labour unions were to socialism. It must therefore be wary of everything that can put it at an abstract remove from that which exists, even legitimately: of the exceedingly logical universal, a panorama whose interest resides entirely in the details of the landscape. At every moment, the general needs to be checked against lived experience: the rawer it is, the more instructive it will be. A good way to come back from the abstract to the concrete, to the *topos*, is by referring to all that is situated and dated: if the historical–geographical case complicates the demonstration, it is very likely to force it to become richer. The reference to space-time will make it come down from heaven to

---

[4]Translator's note: 'Régie Nationale des Usines Renault' was the French carmaker's official name during the years when the company was State-owned, from 1945 to 1990.

earth before the people.[5] Virtue is the sense of the real, which has little to do with its statistical expression: figures and ideas, like photographs, merely freeze one aspect of it in the moment. One must come back to lived experience, to oneself, to one's society: for instance, do not let your love for the proletariat make you forget your own bourgeois life. Do not flee the reality in front of you, even (especially) if it goes against your principles. If it seems bloody and shitty to you, reject degreasers and disinfectants. Do not look down on interests, even your own if need be, as they have an inherent superiority compared to ever-so-virginal ideas: that of existing. Recognize them and, if need be, make them your own, even as you strip them of the beautiful truths that justify them.

Topical utopia could also be defined as spiritualistic empiricism (this in spite of -isms). As a truth revealed, and then demonstrated in an ideology, it describes a global situation, it is a *Weltanschauung*. Its two motivations open its eyes on a world, that of development. Ecological critique is thus at risk of being led astray by the enormity of its task. Over the last few years, a lot of work has been done when it comes to the costs of development and the remedies to be applied. But since it is a phenomenon that encompasses the planet and never stops evolving, everything had to be dealt with at once: science, nuclear energy, the car, demographics, the State, recalibration and so on. The temptation is great to keep to a formula that explains everything once and for all, or, on the contrary, to keep to such and such a sector while forgetting about the whole. There is only one way not to get overwhelmed by everything that claims our attention: to sort out priorities. To distinguish what is essential from what is accidental, the encompassing from the encompassed: the two principles of the ecological movement are sufficiently clear to enable it to do this. There too, ready-made truth, the explanation automatically guaranteed by logic, ought to give way to a feeling for particular cases where judgment appraises.

The priority of priorities, the cause par excellence to which the multitude of effects brings us back, is economic and demographic development. If it is not put into question, we may be able to plug

---

[5] A good way of highlighting the global cost of development would be to produce a yearly balance sheet of the space it consumes in all its forms, and to translate it into maps that would show the public the gradual wasting away of the earth's shrivelling skin under its feet.

a gap, but a hundred others are going to open elsewhere; and local action will be led all the better if we dare to refuse it. In such a vast area, both reflection and action depend on each person's calling. However, aside from development, other priorities dominate. Nuclear energy, especially the spread of atomic weapons, and the exploitation of seabeds, constitute particularly grave threats to life. Let us add to them the industrialization of agriculture when it comes to its quality, the development planning of space-time and the central computer when it comes to freedom. These priorities need to be proclaimed all the more as they are usually kept unheard. They form the basis of any ecological politics.

Our era of change, obsessed with 'concrete results' – meaning those that are translatable into figures and political slogans – has only contempt for that which exists, and especially for whatever makes up the primal and daily joy of living. It is interested at most in what feeds discontent or suffering: see the press and literature. The *écolos* are not above reproach in this respect. Although the raison d'être of their movement is the defence of that which exists par excellence – nature as the bread of life, they tend to be interested in the ideal more than in the reality, in what will be more than in what is. Either that or they pay attention only to whatever motivates their refusal. The author of this book has himself given in to this fault by keeping silent about all that still helps him live in the country. This is understandable: who would think of talking about the air he breathes? True happiness is untalkative, being self-sufficient.

This completely natural indifference to all that feeds the humble joy of being on earth no doubt explains why *écolos* favour organic agriculture at the expense of what is left of agriculture, and solar over more readily achievable energy sources or savings. Ecological reflection and action should on the contrary focus on whatever positive things still exist and are being created: on what remains of nature or of the country or of free human creation. Besides, the positive is, more than ever, tied to the negative, for the former gives its tragic dimension to the latter: today, every oak tree lends itself to pathos because it is destined for the chainsaw, every house is precarious amid the flood of bungalows – every free existence is on reprieve. The needful denunciation of industrial devastation is thus inseparable from the exaltation of all that is its opposite.

The militant *écolo* is duty bound to help the public realize the meaning and beauty of everything that is still beautiful and good: a well-tended timber forest, a truly regrouped grove, a renovated or new house that does not offend its site and so on. If he sees a well-pruned hedge, he will go congratulate its author, and if the work is worth it, he will give him a good bottle. The same goes for all that is worth being devoured by the mouth as well as by the eyes. He will go congratulate the baker who still makes bread, the fruiterer who sells pears and not their spectacle, by taking to him his business and that of his friends instead of turning exclusively to 'organic' food networks.

Above all, his first duty is to pluck and savour all the fruits that remain within his reach. But this is something that is done rather than said.

CHAPTER TWELVE

The ecological community

Being total because it is waged against a totalitarian adversary, ecological action is not carried out on this or that plane, and everyone can find his or her place in it according to his or her temper or calling. It does not entail individual or collective acts, social or political ones, local, national or international ones; it includes all of them or it is nothing. Should one level be missing, action then loses its raison d'être along with its efficacy; and the more basic it is, from the individual to society, and from society to the State, the more needful it is.

The personal basis

The living stone out of which any human association is built is the singular human being: there is no getting around calling him the individual, which is not the opposite, but the condition of the person.[1] The edifice may threaten the sky, but it will only ever be

---

[1] Translator's note: As this book appeared in 1980, Charbonneau also self-published another – his philosophical magnum opus *I Was, an Essay on Freedom*, in which he formulated in similar terms his old argument with mainstream French personalists. There, he pointedly added that 'a few years ago, it was fashionable in some Christian or post-Christian circles to oppose the "human person" whose existence is inscribed in the universe and in history to the egotistical and disembodied individual; which makes it possible to complete the abstract freedoms of Human Rights with the concrete freedoms of the factory and the barracks. For the Human Person was born under Stalin and Pétain,' both of whom were thought of as moving in the right direction by many personalist fellow travellers – often the same ones at different

worth as much as the material out of which it is constituted (*cum statuo*). Especially if we are talking about a society that purports to be based on equality in freedom. One for all (and then all for one), as proclaimed by a formula[2] that is generally interpreted in the sense of the sacrifice of the individual to the collective. People forget that if 'one' is missing, it becomes 'nothing for all'. Hence the fundamentally political importance of personal life, which is individual or else is not. Without it, social life would be lacking in qualities and – as we have seen – the chances of a decisive change would be nil.[3] Freedom – to be and become oneself – is not only a right but the first of duties – no doubt the hardest.

Society begins with the individual, because it must be so, and it is in fact so. Without this carnal and spiritual basis, the ecological movement would only be one more organization and one more ideology. Opposing a state of affairs that practises a strict dichotomy between the individual and the social, private and public life – the mediation being ensured every three or four years by the electoral ritual – ecology must re-establish the continuity between one and the other. This can start at once: private life is but a barren old maid and the one we call public sleeps on the sidewalks. Personal thought and lifestyle do not exclude reflecting and acting on society; quite the opposite. It is in the secret recesses of one's heart of hearts and of family life that are formed, not without toil, the fruits which the public will be able to benefit from. In this area, to the extent that industrial organization and disorganization still allow us, everyone can immediately act, that is, put his principles into practice. Not only will he be able to alter somewhat his natural and human environment, to change his life a little, but he will thus reinforce his personal cohesiveness, work his muscles and not just his biceps, that is, his ability to bear and to overcome contradictions. If he is not able to do it there, I doubt he will be able to in the social and political arena. This way, he will avoid the mistake of that left-wing intelligentsia which, since it leads a life that is (should we add 'concretely'?) bourgeois as it

---

expedient historical junctures. Bernard Charbonneau, *Je fus. Essai sur la liberté*. With an introduction by Daniel Cérézuelle (Bordeaux: Opales, 2000, 34).
[2]Translator's note: '*Un pour tous! Tous pour un!*' – the rallying cry of Alexandre Dumas's *Three Musketeers*.
[3]See Chapter 5.

awaits political revolution, can only conceive the latter in bourgeois terms in its everyday reality.

To act is nothing other than to live. How could politicians who know nothing of private happiness contribute to that of their fellow citizens? Public office ought to be forbidden to people who use it as an escape from their personal problems. And the first condition for thinking and enjoying is to take the time for it: in order to have 'enough', 'too much' is needed. Within the framework of the ecological order we will be talking about, this taking of time, which political action of any value cannot do without, could take the form of sabbaticals and retreats that would be institutionalized for anyone who would take up a public office. It goes less and less without saying that the time for thinking is but that of living; not only to contemplate, but to savour the instant, at the table, taking a walk, in the garden or in front of the workbench. Which should form the ordinary thread of life and not just of leisure. Otherwise, as with some extra-busy executives, leisure becomes just as frenetic as work. They fly on a plane to the antipodes, then suddenly collapse on the sand in the sun.

Ethics is the basis of politics. Ecology, as the science of natural equilibriums and harmony, can only teach an art (and not a science) of human happiness. It must react against a society that reduces delight to its crudest, most violent manifestations. The exaltation of desire, which must be contained in order to be brought to its highest point, is not subject to that dialectic, practised even in ecological circles, that makes one go without transition from the frigid virtues of puritanism to the debauchery of antipuritanism, just as remote-controlled. On that issue, the ancient Greeks would have a lot to teach post-Christians. It is a difficult art, but one that is more needful than ever, especially for Eros, the sun of the senses, and hence of the spirit, that deserves not to be despised by morals or antimorals, that bring it down to the belly's enjoyment. Ecology must loudly proclaim the duty of happiness, which frees from everything that is not man's essential anxiety and suffering and which, having an individual's body and senses as medium, is, above all, personal. Contrary to the assumptions of the moralists and immoralists (or rather antimoralists) who share out our society's allegiances between themselves, being happy is the first of the virtues, without which the others risk being no more than the poisoned fruits of unhappiness. People will retort that happiness is selfish; this is partly true, since it is personal. But what does someone who has nothing to draw from his bank account have to offer others? What is there to sacrifice if nothing is there?

As much as the religion of desire is sterile, the art of tasting life's daily pleasures is the condition of any personal life and work – like the orgiastic feast is that of its confusion in the collective. These pleasures of all kinds cannot be listed here, but for one example to dwell upon as it is of particular interest to *écolos*: the ethics of food. There too, both capitalist industrial standardization and Communist rationing leave us for a while a margin of play – more limited every day – of which it would be stupid not to make use. The same thing goes for Gaster[4] as for Eros, except that, in ecologist circles especially, the suspicion of sin is not completely abolished in the former's case. It is still being accused of vulgarity and selfishness (see *La Grande Bouffe*). And yet, since it too gives life, but on a daily basis, the pleasure of eating, alone or in common, is worth respecting and contemplating; but I do not have a glass of Pomerol at hand to do it while giving a toast for the return of bread and chicken on every Frenchman's table.

Since he may do it for a few more years with minimal effort, the *écolo* must consider as one of his first duties that of eating well (which does not mean too much), and of teaching it to his fellow citizens. Let him stop tolerating that dieticians, economists and governments take his daily bread away in the name of hygiene and morality, profit or the balance of payments. Let him be wary of gurus who trade in natural products that give immortal life, who also want to feed him ersatz. At the table, it is not dreams that must be fed, but the body. Let him refer to his freedom – to his taste, a fantastically sensitive balance with which nature has endowed him, alone capable of detecting the innumerable good or bad tastes of different foods. It will be the first to teach him that savouring is not stuffing oneself and that eating too much is to destroy oneself. Refusing the authority of academia, he will determine his own diet, as his senses almost always take it upon themselves to warn him about what is harmful. His diet will be balanced, not because it is dosed out, but because it is varied. He will practice the golden rule of ascetic delight: 'not much, but good'; if need be, he will reawaken his taste by fasting.

---

[4] Translator's note: Greek for 'stomach'.

Language and ecological reason

Individual existence is a universe that lives and dies with every man. There can be no question of reviewing it here. However, forced to dwell on the continuity of the human fact, from the individual to society, I must remind the reader of the crucial importance of language and reason, even though this issue has been raised in the first pages of this book already. When they deserve their name – which they lose by becoming a medium for lies, language and reason overcome the paradox of the individual in society. For they are at once personal and collective, and if they are not one, they are not the other. Language becomes superfluous in both cases: when an individual is absolutely separated from another or absolutely united to him. Inexpressibly individual or said in advance within the group, as isolation or as collective fusion, silence then reigns. But speech becomes necessary when a singular man has to communicate with his different fellow. Language is thus the prototype and the condition of a society based on freedom. It is not without reason that the religion of the personal God identifies it with the creative Word. A man's mind is embodied and socialized first in the words that have been enriched by the accumulated experience of a nameless multitude.

Thinking through or, to put it differently, weighing one's words, not abusing them and respecting language by avoiding lies as much as possible is the golden rule without which any other form of association is worthless. And this one applies at every moment, for instance as I write these lines. Speech, whether oral or written, as the basis of the social contract, is a means of communication which nature (or Creation for others) has bestowed to man. It would be a betrayal to give it up for audiovisual wizardries that conjure up collective phantasms and gobble up fossil or human energy. Language is a medium that combines all advantages from the viewpoint of nature, freedom and democracy. For zero investment, its possibilities are without compare. If it remains the common language, it is within reach of every individual having a modicum of judgment and education: speech belongs to whoever wants to give one, as TV never will. It is no doubt for these reasons that the industrial system, in a whole variety of ways, encourages the depreciation and the corruption of language; even as the mass of people is losing the habit of reading and writing, specialized literary hacks strive to demonstrate the servitudes it imposes on the mind. The *écolo* must take the opposing view in this area if he wants to do it in others, by striving to give power and dignity back to his speech, which is that of his neighbour. I may point out that a highly ecological means of communication is in the process of being lost: correspondence. How can we prepare the way for a debate or a personal association without it? Between the phone call[5] and printed materials (whose turn will quickly come), there will soon be only individuals isolated in the moment or manipulated masses.

Giving back its value to common language means doing the same for reason, likewise both personal and common, which structures it. This reason, less rigorous but incomparably vaster and more alive than mathematical logic, is the handmaid of freedom, itself at once reason's overlord and its serf. And vulgar reason is democratic because it is available to every man, like speech. In order to discredit it, common sense is termed 'bourgeois': it is the intellectual bourgeoisie that makes sure of this. Whereas common sense, like simplicity, may be the most revolutionary of intellectual virtues: the humility that makes you bow before the enormity of the true or the real, even if it makes you look like an imbecile. The inability of our society's intellectual leaders to take note of the effects of development on a finite planet is but a total lack of common sense, as is their tendency to judge political facts in terms of an ideology.

We need to appreciate again the reason expressed in common language, not only to restore democratic knowledge but also because it often expresses what is essential, whereas erudite jargon and logic only communicate what is incidental and what is special. In their public relations, *écolos* would do well to favour arguments that can be stated in the language of everyone while avoiding those that belong to a science – or, needless to say, an ideology – that is but magic for the ignorant. This way, the people is given back its dignity and is taught anew to think for itself, even as we get straight to the point. Thus, when it comes to nuclear matters, we would do well to dwell on the consequences that do not belong solely to

---

[5]Translator's note: In an untranslatable play on words, Charbonneau uses scare quotes in *"coup' de téléphone'* to underscore the loud, aggressive, isolated nature of the phone 'shot', which a phone call translates as in French, on the pattern of a gunshot.

discussions between biophysicists. And these are the most important ones. Namely, a police society, the concentration of knowledge and powers, the control and repression imposed by pollution and the risks of accidents or terrorism. To this may be added the nuclear plant's inscription in space with its cooling towers and its very high voltage power lines. As far as the issue of financial and energy cost effectiveness, if that discussion belongs to experts, everybody can readily understand that they do not always agree; and that, before producing megawatts and billions tomorrow, today we will have to invest some, that is, to sink them. Besides, there is nothing to stop us from enhancing common sense arguments with others borrowed from reliable scientific vulgarization.

Ecological society and meetings

There is no human life or action without a society; our very nature is social. Whoever has truly experienced his solitary freedom knows to what extent identification with a group is an easy way out as much as a duty. But from the family to the empire and the species, society is innumerable; and until now, one could belong at once to a township, a nation and a church. It is this manifold belonging that contributed to the richness of persons and to their freedom. And yet, perhaps even more than the diversity of individuals, it is social plurality that tends to disappear, absorbed as it is in a total society under State control. One of the tasks of ecology (let us add social ecology to sound socio-logical) is to restore it. Let us recall its principle: *Small is beautiful*.[6] If it is not just a slogan, it implies not only the priority of the person over the group, but also that of the small society over the large one. Because it is closer to its environment, its management is simpler and interpersonal relations are easier. If the ecological revolution does not change the personal or social microcosm, it will fail to change anything in the macrocosm.

Hence the necessity for the ecological movement to participate in the defence committees that keep its contact with the basis: nature, everyday life, local populations. Without such a return to this

---

[6]Translator's note: In English in the original.

experience of a basis that is spiritual because it is material, the ecological movement will be nothing but one more ideology and party. On the other hand, if this contact is kept, ecological politics will no longer be the alibi, but the natural extension of what exists and what is essential. And with better results than Marxism, which has merely politicized labour unions, ecology could bring to defence committees the general view that would justify their action; it would help them widen their often-limited horizon, overcome rivalries between petty bosses and federate.

Tomorrow begins today. The ecological revolution is also happening here and now in all kinds of microsocieties, among which the most demanding ones attempt to live up to a utopian model. *Between* man and nature, and especially between man and man, they are trying to establish new ties that would not be ones of coercion and necessity but those of love and freedom, and thus of equality between individuals. The community (distinguished here from purely social society) based on a contract entered into between free and equal individuals is not given at the beginning as Rousseau believed, but is a project to be realized against all the odds stacked by the universe, through the sole power of the loftiest human desire. Turning the ideal of 1789 into a reality is one of the ecological movement's reasons for being. An enterprise whose paradoxical nature becomes glaringly apparent from the very start of communities. Even within a free and spontaneous human-scale association, there is a need for an institutional minimum: of truths, rules and disciplines to which one must submit. A community cannot be instituted on the basis of pure nature and pure freedom. It is not established in heaven but on earth, somewhere in the France of 1990. How do we integrate within the established order with which we want to break without making compromises? Clearly, the community's external policy is just as problematic as its internal politics. How do we realize the paradox of freedom within nature and equality within one's society, and within the opposing society that can only swallow you up, whether it is with or without violence? There is no ready answer available; it is up to those involved to provide the answers as their cases require. But the minimal condition for success is to know this: to be very aware of the paradoxical nature of the enterprise one is launching upon.

There can be no more ecological political action without a social base than social life without personal life. And this social basis can only exist at all levels, not just the local one but the continental and planetary ones as well. Plural society (let us not call it plural-istic, for it is to be found much more in facts than in principles, as has been the case until now) calls for federation, a kind of national or international ecosystem combining a multiplicity of societies that could not exist without it. But the federation is not necessarily the State, it can take many forms, for instance that of an order or an international fraternity.

The ecological movement's raison d'être is to promote a society that does not do violence to nature and to man, where the relation of the latter to the former and of man to man would be restored. The first place where this ecological society should be instituted, it would seem, is the movement that claims to be ecological. Unfortunately, it is often better described as ecologist for, as a victim of its own overhasty success, it is sometimes hardly ecological in its means. Fooled by the mirage of power that seems so close even though it is far in the distance, the leaders who swear by *small is beautiful*[7] unwittingly look down on the small and dream of ruling over the masses and the State. Forced to act within the massive and centralized framework of the opposing society, ecology's worthies are tempted to borrow its means. Succumbing to its pseudo-facilitation of transportation and communications, they run just like other politicians or businesspeople from one end of Europe to the other, without taking any time for reflection or exchange. Caught up in external or internal struggles for power, they are only interested in what allows them to reach it, or gives them that illusion. What they say or think they are doing is organized with a view to the major media they in no way control: ten lines in *Le Monde* or five minutes on TV become the goal that determines all the rest.

As they grow older, ecologists, in turn, practice the dichotomy of ends and means and figure they can transform society by acting according to its methods, whereas it is essentially defined by its techniques: to accept them usually means being swallowed up by them. In many cases, the ecological meeting has nothing ecological about it. As soon as it grows beyond the size of a small group of buddies, it remains caught within a dilemma from which it seems

---

[7]Translator's note: In English in the original.

unable to get out. Either it remains spontaneous and democratic – but then it is inorganic and gets lost in endless discussions that never come to anything. Or it organizes, but according to pre-existing procedures that it does not question, namely the colloquium or the congress, the typical meetings of neo-industrial society. If it wants to change anything, ecology must start at the beginning by giving some kind of ecological character to ecological meetings. Otherwise it will evaporate into words instead of translating into something that actually exists. Here again, the ecological movement must open its own path between infantile idealism and senile realism, by inventing forms of meetings and communications fitted to the ends it pursues.

When established ecologists meet in any kind of official setting, the meeting takes the form of a colloquium (ironically so named). In principle it is about getting information and discussing, but it is actually a ritual of integration and promotion within industrial society. It is set up by an organizer with a gift for public relations[8] and funding (especially if simultaneous (?) translation needs to be launched). The value of a colloquium depends on the number of worthies among participants; if they are many, they will all need to be there and the ceremony's success will be assured, *Le Monde* will devote an article to it and maybe TV reporters will come. Hence the abundance of presentations: a dozen of fifteen minutes at the most. How do you follow such a marathon to the end? A dubiously competent but particularly well-known public figure comes to do his number and then goes, having no time to waste. Such a high mass is only natural for someone who is used to it, but the outsider notices the frequently stereotyped tone, the superficial character of the presentations, often limited to ultra-specialized detail, the caution of professional speakers whose social status or career is at stake. In principle there are discussions but, since time is limited, they must be cut short. From the platform's empyrean height (is it an Olympus or a shooting gallery seen from the floor?), truths fall on a public that keeps its critical impressions to itself. At most, it is invited to take the mike. This ritual object, a kind of aspersorium for the sprinkling of blessed words, becomes so indispensable that it is used in ecological meetings of twenty people endowed with

---

[8]Translator's note: In English in the original.

normal voices. Keeping it or capturing it is a sign of power. Most of the times it falls into the hands of loudmouths with a knack for sloganeering.

The congress is related to the colloquium, only less hushed. Its organization also makes it a ceremony: a mass if it is that of a monolithic party, a match if it is split between currents as is the case with ecology, in any case a spectacle heightened by the media. No more than at a colloquium is there any information or discussion: it is just a struggle for prestige and power between personalities and currents. And there too, the only place where people meet and sometimes discuss is the corner bistro. Save for certain commission meetings, what might have been a clash of personalities becomes a battle between groups and their leaders; the experience of one hardly enriches the other's, and divisions are exacerbated to the enemy's greatest benefit. At best, the conclusion takes the shape of a murky compromise in which twenty people get together to change a comma. This conclusion is also established with a view to the media whose lines and time are counted, and who demand it be expressed in their lingo. It is easy to imagine what a dog's breakfast this gives; fortunately, there is only one or two spoonfuls' worth.

It may be that this kind of publicity show[9] is necessary to spread ecology (and especially its slogans) among the public, but it will have a minimum of content only if it is prepared by genuine meetings where informed or stake-holding individuals will have compared information and viewpoints. For one may well ask if all these colloquia or congresses are not chiefly meant to exorcize the difficulty of communicating in our society that is at once too large and too complex so that, for lack of time and criteria, by their sheer superabundance, relations and information become impossible. Especially if the 'meeting', taking on an international size, makes use of this deep-frozen speech: simultaneous translation. Failing to translate anything, it clearly has but one function: to magically negate the curse of Babel.

If the ecological meeting is not ecological, its fruits will not be either. It is thus of prime importance for the movement that claims to be ecological to ponder the costs of the means specific to the society

---

[9]Translator's note: In English in the original.

it condemns when it uses them. And above all, to create specifically ecological forms of meetings, communication and events. The May revolt provides some examples, albeit too spectacular, of this effort of imagination (parallel press, clandestine radios, the style of certain demonstrations).

The ecological movement lacks a counter-institution that would enable it to hold a meeting worthy of the name. It presupposes a limited number of participants (let us keep trotting out *small is beautiful*[10] – except when it comes to electoral results!). Hence a selection, which should not be done with the sole criteria of notability. To allow for real debate, one would seek to eliminate all power play, for example, by refraining from all media publicity. The work program would be established in advance through contacts and correspondence between prospective participants. Like the meeting, its venue would be ecological, located somewhere deep in the countryside; and the style of the meeting, simple without austerity, would also be in line with ecology.[11] We must forbid those plasticized luxury canteens for international colloquia where the impersonality of comfort is matched only by the insipidity of the food. The meeting would be sufficiently long to guarantee a few days of common life and the program would be loose enough to allow room for reflection, for personal relations and for pleasures, as is the case in ordinary life. Without being timed and specialized, it would be centred on this or that essential or urgent theme of ecology so as to prevent the discussion from wandering. It could also be the occasion of a balance sheet of the general situation and of a self-critique of the movement, which can be done between individuals who are close and relaxed without degenerating into misunderstandings and conflicts. Should it happen that the decision was made to conclude with a public declaration or act, once the issue was debated between persons, the good rule would seem to be anonymity under a collective banner.

---

[10]Translator's note: In English in the original.
[11]Translator's note: This was just the format of the meetings of Bernard Charbonneau's southwestern splinter of the personalist movement after his break with Emmanuel Mounier in 1937, issuing in part from a similar critique of the national congresses of the latter's *Esprit* review.

The ecological order

The organization of this type of meeting would be very suited to an ecological society: let us call it an order, and should this not be to your liking, a club. It is just as necessary to political ecology as the Fabian Club has been to the Labour Party. The ecological current will be able to spread among the public only if it gives itself the means of remaining itself, failing which, even as it feels its reach expanding, it will dissolve in said public. Ecology will take a political form only if an institution maintains its spirit and praxis. Vis-à-vis the ecological party, it would fulfil the function of an authority facing a power in the manner of a church facing a State, though these two terms should be somewhat qualified. It is the minimum of 'separation of powers' without which there is no society: in Montesquieu's time, there could be no question of calling for it, as it went without saying. Monastic orders and freemasonry are pretty good examples of this kind of association, on the condition of innovating so as to avoid the mistakes that have led them to degenerate. To underline its mistrust of ideology – if not of the idea, this order or club would choose an ecological symbol that is as concrete as possible: everywhere cut down, at best wrapped in cotton wool by the *Raubwirtschaft*,[12] the Tree seems particularly rich in meanings.[13] To the extent that the encompassing society leaves the association enough means and rights, the latter would become the prototype of a future society that is respectful of nature and of freedom. But it would not be an ideal society, just a very real one, which would take into account the possibilities of the situation and of its members. Just as with food 'not much but good' is a good formula, it is better to set oneself a modest goal and reach it than to go from illusion to disillusionment.

---

[12]Translator's note: German word meaning 'plunder economy'.
[13]Translator's note: Around the time of writing, Bernard Charbonneau launched the Groupe du Chêne, named after the oak tree behind his Béarn country home, at whose foot this ecologist club would meet and where he now lies buried (together with his wife Henriette). Among Charbonneau's handwritten notes of the late 1930s can already be found much more detailed blueprints for an 'ecological' order of this kind, as the spiritual authority (based on hands-on experience of self-reliant low-tech survival skills wandering the roads of unfamiliar rural areas) of a new personalist society envisioned as arising in southwestern France.

Although its participants would necessarily be chosen by co-optation, it would be preferable not to turn it into a collection of billboard worthies who generally give their names and nothing more: that would go without saying since this would not be about gaining media attention. This social ecosystem would itself be all the richer for mixing diverse elements: intellectuals, militants, people of all trades and private persons responsible for reminding those in charge of the daily situation of the base. Its members would pledge themselves to a minimum of commitments in keeping with the rule established in common. A financial commitment of course, not so much to ensure funding as because it sums up the others: this is often the reason why people refuse it. Among these commitments, let us mention those to meet, to visit and to correspond, some members having a more specific vocation to maintain contact, like the wandering monks of old. Let us add to this for those in charge the return to the base and a time of retreat. The rule (still with a view to the minimum) would establish a kind of deontology of the *écolo* in industrial life; in some extreme cases, some political memberships or positions (in the banking system, advertising or the police, etc.) would automatically result in exclusion. As for rites, feasts or ceremonies that would symbolize the association, no need to define them in advance. That would mean granting more importance to forms than to substance and mistaking the effect for the cause. We may rest assured that the day when this society would exist, these signs of life would be given to it additionally.

The function of an authority is not power, but reflection, information and communication, judgment. God forbid this institution should start to look like a power centre. Its business is to provide for meetings in the sense sketched above, the constitution of documents, of a library and of an ecological directory that would allow *écolos* to get in touch. It could also devote itself to the edition of a serious review, of which there is need in a fledgling movement that only has at its disposal publications that forbid exceedingly long feature articles. It is only on exceptional occasions that the institution would manifest itself in a public and anonymous declaration about such and such an event or act, especially the kind that official news blackout. But discretion would really be the norm. Without being secret, this kind of society should refuse advertising and spectacle, having no reason to establish relations with the media, since its business would not be the mobilization of the masses.

Its natural organization is federation. That is a basis of cells rooted in their setting: village or regional committees, township or neighbourhood clubs. The need to keep them on a human scale (let us recall S.I.B.[14]) forces us to consider a *numerus clausus* beyond which another group should be founded. These local societies would be grouped like others at the regional, national and international levels, where local leaders would have the opportunity to broaden their horizon, to know each other, to compare their experiences and to coordinate their action. Of course, it is out of the question that by broadening their horizon these leaders should lose sight of their roots and their base. From time to time, immersing themselves in the land again will help them.

In the case of the nations of Europe, we must insist on the necessity of a European federation. Ecological problems are local or continental, more rarely national; especially in a narrow and overexploited continent, they cannot be regulated at the national level (see the pollution of the Rhine and the English Channel, the problems of the Alpine or Pyrenean mountains, etc.). They are the same ones that arise for all peoples from different angles: the defence of the Périgord countryside or of the Aquitaine coast does not matter only to Aquitanians, but also to the Germans and Batavians who come there to revive in the summer. If the optics are different, viewpoints complement each other for that very reason: someone from Switzerland or Belgium will rightly emphasize the damage inflicted by private enterprise and provincial governments, whereas a Frenchman will insist on the ravages of the central State and of town and country planning. They could thus all teach each other by showing how private capitalism and the State combine and will keep combining more and more.

Let us leave it at that; a book cannot sum up a world, even if it is only a microcosm. Everything in its own time. If we succeed in laying these foundation stones, we will not have lost ours, and the rest will ensue very naturally – ecologically.

---

[14]Translator's note: *Small Is Beautiful*.

CHAPTER THIRTEEN

Ecological politics

Whether one is subjected to it or imposes it, there is no escaping power: the only way to do it is to exercise it over oneself. Outside of power, there is only one thing: putting up. To refuse this would be to refuse to live. It is therefore natural. But man adds to it not only the awareness, but also the enjoyment of wielding it – whatever its goal or its pretext may be. He savours the delight of mastering the object or the subject that resists him, sometimes to the point of sadism; either that or the delight of submitting to the power that dispenses orders and blows. Power is male, especially when it starts with a capital letter; and it is doubly so when it does not wear trousers, but a cassock or a skirt, like Catherine the Great.[1] This is why the highest lesson that man in the masculine can receive from man in the feminine is indifference to power and the ability to love, a tie that is the opposite of the iron chains forged by engineers and tyrants.

Ecology and power

But more and more, all powers come down to one: Power, or in other words the State that organizes social powers that used to add up individual powers. And if gathering to act on nature and men is at the same time enslaving oneself, from now on we are slaves to a single all-powerful entity on which our happiness or unhappiness

---

[1] Translator's note: Charbonneau puts the adjective for 'Great' in the masculine to underscore his point.

depends. We should therefore not be surprised if the religious divine has been succeeded by the political divine. 'Everything is political' – it is even as we combine the delights of the master and the slave that we rehash this formula. Politics (do not ask me which, I know it since I do it) is Action that encompasses other actions. The ability to flatten everything out to just the political dimension remains typical of the age. And if experience forces us to discover other dimensions, we conclude, 'Nothing is political,' whereas this all or nothing is the sign of an unbalanced time and spirit from which ecology should free us. Going beyond the false dilemma that opposes anarchistic communities to leftist militants, we must recall that one does not change a regime without changing life, and one does not change life without regime change; the task is vast enough for all *écolos* to contribute to it according to their abilities. More than ever, the exercise of political power is one of the necessities of the human condition and of human action. But if it is not nothing, it is not everything; saying that we must allow for it means recalling by the same token that its realm is – and must be – limited.

Politics in the true sense of the word is the management of common affairs by citizens in a sense that they will deem positive. Handling it in such a way that it no longer be a nihilism of acting for acting's sake as justified by some abstract ideology – an agitation that turns to a mania like that for tobacco, that it take into account the concrete environment and the ends to be pursued, that it be done by individuals and not by specialists of the manipulation of the masses, is only giving back a meaning and content to politics.

In the West, a more or less explicit two-party system fulfils the function of the one-party system in the East. There, political gatherings take the form of the Right–Left game: if the teams do not wear the same jersey, they face each other on the same field, outside of which nothing happens. This for–against that gathers the people, a legacy from another era, becomes even more gratuitous than the opposition of the Blues and the Greens in Byzantium,[2] in the face of the issues of the year 2000. If ecology wants to change anything in politics – and hence in society – it will have no choice but to get out of the game, and thus of the camp to which it is sociologically related. As we all know, 'neither Right nor Left' is a

---

[2] Translator's note: The two dominant colour-coded stables in the horse races of the Hippodrome, whose rivalry polarized Byzantine society and channelled its tensions.

right-wing formula[3] – since it is not of the Left. And yet, the refusal of it is evidence of perfect immobilism, and *écolos* are condemned to give it a content if they do not want to disappear as an original movement.

For two centuries, the Right and the Left have shared out between them the whole of opinion and values. On the Right is reality: nature and necessity, tradition and the established order. On the Left, even when it claims to be materialist, the ideal: every liberty for all, the individual, justice, progress and socialism. On the Left as on the Right, the least critical mind soon discovers that these opposite values are but weapons to be thrown in each other's face in battle. Besides, as in a busbar, it is enough for the situation to change for the two camps to exchange them, the Right becoming pacifist when the Left becomes patriotic. But this kind of opportunistic conversion only goes to show how little attachment the Right and the Left have for their own principles. Thus, in turning progressive, the Right has hardly discovered the values of the Left; it has merely betrayed its own. And for its part, by coming around to family values and patriotism, the Communist Party has done nothing but recuperate to other ends their lie and their use by the pre-war nationalist Right. There is something worse than the reactionary or the progressive: it is the progressive reactionary, or the reactionary progressive who, like national–socialism, adds up and confuses the lie of the Right and the lie of the Left. In France, you only need to look on the Right and the Left for the two parties that are best at performing this manoeuvre.

We must therefore resign ourselves to get out of this field – which is not easily done, for this means getting out of the society to which we belong in having either right-wing or left-wing guts. But having

---

[3] Translator's note: This assumption was lampooned in 1966 by Jacques Ellul in *L'Exégèse des nouveaux lieux communs. A Critique of the New Commonplaces* (New York: Knopf, 1968; Eugene, OR: Wipf and Stock, 2012), being the standard progressive objection to the nonconformist stance towards the customary political spectrum claimed by a minority of spiritualist intellectuals – personalists in particular – in 1930s France. It was in that context that Charbonneau and Ellul developed this 'new research method' of 'exegesis of commonplaces' to deconstruct media discourses, twenty years before Marshall McLuhan in *The Mechanical Bride* (New York: Vanguard Press, 1951) and Roland Barthes in *Mythologies* (Paris: Editions du Seuil, 1957). The 'neither Right, nor Left' slogan would be revived by France's Green Party in 1986.

done this self-critique, it turns out that in many cases, if we go back to their origins, the values of the Right and the Left, far from being opposite, are complementary. It is the recognition of the demands of nature and of society, the teaching of the past that makes of the ideal of freedom and equality something more than a dream. This overcoming and this rejuvenation of values that had heretofore been opposites have been undertaken by ecology in its emphasis on nature against 'progress'. It only needs to continue by staying the course between its right wing and its left wing (without one or the other, one goes around in circles). Why should it be either backward-looking or progressive? Between a past that was hardly rosy and a future that is nothing yet, its business is the present where everything exists and is decided, yesterday's knowledge and tomorrow's forecast being helpful in this respect. Just as the ecosystem is the putting together of differences, ecology is the synthesis of partial (in both senses) truths that militant polemics set up as opposites.

Its first task is the defence of what exists: this is a great change in a society that keeps on repeating that you must *a-dapt*. What a progress it would be if the ecological revolution gave us the time to breathe! It is revolutionary because it is conservative, defence being but the first phase of attack. The protection of all that subsists – endangered nature, sites and social categories – quite naturally prevails over the achievement of new energies that will come later. Thus, for the rural habitat, ecology's task is less that of prematurely inventing a type of house that is liable to making the land ugly in its turn than to protect what is left of the existing heritage whose roofs crumble every day. As for the rural habitat of tomorrow turned today, it will grow on its own in a style of its own if social conditions change along with their pace. No need to spell out how much passion and invention are demanded by these defence actions inspired by the spirit of offensive; *écolos* know it all too well.

The creation of a different society does not exclude the defence of the interests of various social categories that are threatened by change; on the contrary, it is one of the movement's defining priorities and conditions for success. If it practices it for real, that is, without deceiving itself about the virtues of the average peasant, grocer or craftsman, nor deceiving these backward people to pick up their votes through demagoguery. This niche is yet to be occupied, and it is huge, since development simultaneously enriches and impoverishes nearly all social categories. If S.I.B. (forgive the acronym[4] that must be recalled every moment), sympathy for small peasants ought to extend to retail and crafts that also deserve to be preserved and perfected. There again, before soaring to heaven, the imagination has the opportunity to practice with numerous specific local cases. In our society, the handicapped are not the only ones to be so; the number of potential allies of the *écolo* movement is considerable and far exceeds that of its voters. There are the old people (hence the young who are future old people) who wake up in a world that is no longer theirs, the hunters and fishers whose economic boom destroys their raison d'être by suggesting to the wealthy by way of compensation that they go kill half a world away the hare they used to have in front of their house. It is silly to condemn them when they have a vital interest in conserving the natural environment: provided they are not poachers or slaughterers of farmed game. It is better to denounce the destruction of all wildlife by pollution, roads and the organization of leisure – including various safaris – in the name of hunting and fishing that is worthy of the name.

Today, as yesterday, revolution is inseparable from tradition (see the Swiss and French revolutions that started out as a defence of custom against monarchical centralization). Because the principle of the established disorder is today the plunder of what exists, its defence is revolutionary: anyone who insists on saving what is left of rivers or streets learns quickly at his own expense that, within a legal framework, one can at most conduct delaying actions. 'It's at least something', he tells himself as consolation. If he goes any further, he runs into a wall, be it disregard for laws or new laws that abolish them. So, to a society that knows only facts, he is tempted to oppose a fact: sabotage and terrorism. But by answering violence with violence, he places himself on its turf and, in his turn, risks being seized by a rage driven this time by impotence. He makes up for his scant means by using the most violent ones if he finds the help of a foreign organization. He becomes a specialist of murder, explosives and hostage takings, a spy and a hired killer who ends up forgetting why he became that. Through terror, he exploits in the masses the lowest feeling: fear. And between him and the official police, the winner will always be the one who will have been the

---

[4]Translator's note: For *Small Is Beautiful.*

most successful in instilling the fear of being killed. At this game, the official terrorist, if he does not hesitate too, has a good chance of winning. But if his enemy brother, supported from abroad or by circumstances, comes to power, the practice of secrecy and murder will make him an even stronger policeman.

How do you answer force by force when every relationship is reduced to the latter – and you are the weaker one? The question that clandestine or State terrorism puts to any minority movement deserves to be decided in the negative, or at least approached with fear and trembling: But are there any terrorists with trembling hands? If not, we risk being taken in by the irresistible urge to wield power in our turn more than to save the earth and freedom. There again, between a legalism that condemns us to conformity and impotence and a terrorism that leads only to bloody failure anyway, the ecological movement is obliged to refuse the dilemma in which the present system would imprison it. Hence the necessity of using all legal means to use all the gaps or contradictions that persist in our society without stopping at that. As for internal or external war, it is a last resort, forbidden to the weak and without rules, of which it is ridiculous to speak as long as necessity has not driven you to it. To each day its own task, the day after tomorrow we will see about playing that most gripping of games: little soldiers.

There is no more magic key to open the door of this wall one ends up hitting against than there is any dynamite to force it open. At most, we can say which keys are false. But he who does not seek never finds; on this wall too, one must first break one's head to open a way out. Let us not forget that it is also within us: in our powerlessness to conceive anything other than what is given.

Elements of ecological tactics in politics

Its strategy being reducible to seizing power, politics, as it is usually understood, is limited to tactics; but since the latter depends on circumstances, it is rather pointless to theorize it in detail. The art, as always, consists in asking oneself questions and not letting oneself be caught in false dilemmas. Thus, need we act within the framework of the official government, opposition and unions or do we create our own? Why not both one and the other? Why wouldn't *écolos* take part in this ministerial action or that campaign of the CFDT or the Communist Party? It all depends on what it will be worth; the answer cannot be given in advance. And it is precisely insofar as ecology will know what it is and what it wants that it will be in a position to venture there without getting lost. The ecological movement is not called upon to take a stand vis-à-vis the government or the parties as such, but to take decisions according to their actions, rather non-existent as they are.

In many cases, it is not so much participation in official activities that should be incriminating as the spirit and the way in which it is practised. It is generally the province of born collaborators whose temper drives them to flee conflicts and fights, as that of some others drives them to seek these things without motive. Their reason is less the service of ecology as a need, justified by blinkered efficiency, to get closer to power. To remain in their official position, these people are prepared to say *amen* on any occasion or to be silent. Whereas the only excuse for being there would be to manifest a quirky presence by saying what is left unsaid and to accept what is offered, be it a subsidy or a natural reserve, without mincing words about what is denied or the explicit or implicit conditions these gifts come with. Whoever is able to do the most is able to do the least: one can accept alms while continuing to demand the rest of what is owed. Unless one practices the accelerationist politics of hastening breakdown,[5] like some fanatics, which is the opposite of ecology. There is every reason to be sceptical about the benefits to be derived from this collaboration: one needs as long a spoon to dine with the State or with the corporation as with the devil. Quid pro quo, the ecologist must not forget that by his presence alone he gives a green label that is worth 3 to 5 per cent of the vote and, so, more than fine words. Consequently, let him be wary of those alibi commissions and sessions where chlorophyll-laced chatter serves as cover for destructive development. One of the advantages offered by participation in official actions is the ability to noisily resign or, better yet, to get thrown out.

Since the possibilities of acting within the big parties, solidly organized and held under control, are so limited, and we cannot

---

[5] Translator's note: '*la politique du pire*', a French idiom that translates literally as 'the politics of the worst'.

count on the unions, not even the CFDT, the moment job-creating development is involved, the establishment of a political party explicitly recognized as such in the service of ecology seems inevitable. On the condition that ecology not be at its service. If it is founded, it will be able to call itself ecological only by respecting a minimum of rules, like the ones that Europe Écologie gave itself during the last elections, by establishing a murky compromise (this being a tautology) that somehow reconciles nature and freedom with organization.[6] Nothing would forbid that it sign an agreement with other parties, on the condition that it be on concrete points, not on myths or vague principles. A gathering of all marginal movements of extremely diverse and questionable hues should never allow us to forget the ends that are peculiar to ecology. The cases of refugees, disabled children, battered women, homosexuals or common law prisoners deserve attention, but they are not the business of a party whose essential motive is to save what is left of earth and freedom.

Let us not dwell on the problems the ecological party will have to discuss the day it exists. How do you find money without getting bought or depriving of it the rest of ecology? How do you win over the public otherwise than through the myths and lies of a propaganda: how do you tell it we can no more have the benefits of ecological equilibrium without its costs than development without its own? At this game, we will have to resign ourselves to losing, as much at the level of ecologist orthodoxy as at the level of returns. Above all, just as the State is the State, the party (which keeps its small p) is only the party. It fulfils its own function, not any other; if everything came down to it, there would no longer be any movement. This is why it is not a bad thing that, at their own risks, *écolos* wander to the Left and Right or withdraw to the land. Should they err and compromise the movement, one of the tasks of the ecological order discussed above would be to make use on such exceptional occasions of an independent authority to solemnly state its considered opinion.

---

[6]Translator's note: Charbonneau is referring to the Europe Écologie list of Green candidates put together in France for the first European elections of 1979, not to be confused with the French coalitions of the same name in the European elections of 2009 and 2014, whose merger with the national Green Party (Les Verts) in 2010 led it to adopt its current name Europe Écologie Les Verts.

A non-economist economic policy

What follows is the sketch of an ecological program, indispensable for whoever is thinking of a future that is not that of plans, always the same and short-term. A trifling sketch for someone who is without power and is talking to a movement that has a long way to go before getting in government. But these distant milestones are just as necessary for the one who wants to stay his course today already. Since reviewing everything from A to Z is out of the question, it will suffice to point out a few in the key area of economics, agriculture and leisure.

What matters to ecology more than any gadget meant to exorcize questions as much as to settle them is a policy – a general one, needless to say. It is not solar, but the savings and the diversification of energy to which it contributes, not 'natural' products or 'organic' agriculture, but agriculture, period. Not only soft technologies, but an economic policy whose principle along with certain modalities have already been mentioned.

The material basis of everything is economics. The reader may be surprised to find this formula coming from the pen of an adversary of the neo-industrial system. Another paradox to maintain: it is the one who does not want to be the plaything of economics as of politics who must devote all his attention to them. The religion of economic facts goes along very well with idealism. 'Stewardship will follow' has been the vain and naïve formula of a regime where everything hobbled along behind a stewardship without a steward.[7] The current disorder calls for a view of economics that would not be just the province of economists, and the questioning of its twin pillars: Development at all costs and the Market and, even more, their explosive combination by the State.

The ecological movement counters growth with terms such as development – which is supposedly not growth, 'ecodevelopment', at most 'zero growth'. They all go to show how much trouble it has

---

[7]Translator's note: Charbonneau is referring to De Gaulle's alleged (but spurious) appropriation, upon coming back to power and ushering in the modernizing Fifth Republic, of Napoleon's famous quip '*L'intendance suivra*', the conqueror's answer to concerns about the logistics of the supply corps during his relentless lightning campaigns. In a civilian context, the phrase expresses the subordination of economics and management to the political goals defined in all sovereignty by the French State.

getting beyond our era's prejudice. Even zero growth is a negative formula where growth remains the criterion. One might as well say 'zero degrowth'. For if the crisis continues, the maintenance of production at current levels may well be considered a splendid success before long. It would thus be better to speak of *equilibrium*, the ecological term par excellence. This pursuit of economic equilibrium is especially urgent, after many decades of uncontrolled growth, if we want to sort out its humanly positive gains. Otherwise, when it comes to knowledge of its effects, future planning will only be able to forever run after a change that will have changed once the future has become the present.

The moratorium, a pause conducive to scrutiny and reflection, is a good formula that is not only good for nuclear power, it can apply to all cases in which a limit has been reached as well as to development in general. Thus, for the Côte d'Azur or the Vendée, a moratorium on tourism development would give the time to set up urgent sanitation equipment instead of inviting hundreds of thousands of new summer holidaymakers to settle in housing that has no toilets.[8] A moratorium on energy development, even as it would spare the supply of fossil wealth and limit pollution, would make it possible to carry out the development of new renewable energy sources. Otherwise we will be acting in the heat of the moment with all the risks of error that this entails. But we should not fall back into the error of 'nuclear everything' with a 'solar everything' which, used uncritically, would lead to the invasion of an incompressible and fragile space. Once again, the solution will not come from these or those technical means, but from their combination and from energy saving – the first source of energy, according to Mr Bosquet,[9] and the first condition for which is the slowing down of development.

---

[8] Anyone who is familiar with the cement wall – nay, the cement cliff of the Vendée coast – can only wonder at a plan that calls for 187,000 more beds. Depth land-use planning will be carried out. It figures: aside from a few forest squares, the coastline is full.

[9] Translator's note: Michel Bosquet was the pseudonym under which ecosocialist philosopher André Gorz, who originated the term *décroissance* ('degrowth') used here by Charbonneau, co-founded the left-leaning news magazine *Le Nouvel Observateur* in 1964. He was eventually sidelined from its economic pages a decade later, as his crusade against the nuclear industry cost it advertising revenue from Électricité de France.

Just as an economic development policy assumes demographic development, an economic equilibrium policy assumes demographic equilibrium. In the Sauvy-Debré system,[10] more and more young people are needed to increase production, and more and more production is needed to provide jobs to young people; even more assets are needed to pay seniors' pensions, but since young people are future seniors and children are inactive, more and more ... Fido runs faster and faster after his tail. This hellish spinning will have to come to an end someday, in particular in developed countries where overpopulation goes along with overconsumption. Since demographic movement tends to equilibrium on its own, it will likely be enough to let nature take its course by suppressing the artificial incentives that allow the State to increase the birth rate. But this raises many issues, among others that of an agreement between nation states.

The other pillar of the temple is the Market, justified in the West in the name of the laws of science as Marxism is in the East where, incidentally, it is combined with the Plan in more or less official forms. Profit and the accumulation of capital are but elements of a generalized 'value creation',[11] thanks to money and the calculators that put a figure on any free good, whether natural or human, for the purposes of exchange on the market. But it is not the village market we are talking about, and there, advertising often allows offer to determine demand. Everything that used to be freely offered is swept up in it; it will soon be the turn of the air we breathe. The market mechanism identified with the economy by way of the play of prices and competition is the engine of development, even in the East, where the State has to take into account the profitability of production. By selecting the most competitive methods and

---

[10]Translator's note: Charbonneau is apparently referring to the book by Robert Debré and Alfred Sauvy, *Des Français pour la France (le problème de la Population)* (Paris: Gallimard, 1946), a post-war natalist call for 'more Frenchmen for France'.

[11]Translator's note: Charbonneau talks of '*mise en valeur*', a term for 'development' or 'promotion' that literally translates as 'putting/turning (something concrete) into (abstract) value', more than 'value creation' (out of 'nothing' as it were). This process is comparable to the enframing of any given reality as resource for evaluation by technique as *Gestell* in Martin Heidegger's thought. It is significantly on the same pattern as this commonplace French idiom he underlines here that Charbonneau will talk of '*mise en science*' in *Ultima ratio*, his critique of the related imperialism of modern science, included in *Nuit et jour. Science et Culture* (Paris: Economica, 1991).

businesses, the market drives an ever-greater, thorough exploitation of earth and men, and organization, that is, standardization and concentration: in a word, Progress.

By thus extending to the oecumene and to a politics that does no more than translate its demands, the market fashions life and its environment even more efficiently than would a totalitarian will. But no more than communism does that will know what it is doing, especially since it has lost its head and multinationals have become impersonal administrations. For, like communism, economics is less a science than a religion – a sacrificial one of course. Its current priests may make adjustments by combining dirigisme and liberalism so as to avoid a crisis of the same kind as that of 1929; they are merely taking a step back for a better jump into more serious ones like the energy crisis, due to other blind spots of economic science. Econometrics may multiply figures to give itself an air of rigor, but it never steps out of the closed world it has fabricated for itself. For if it is true that there can be no policy without an economic basis, there can be economic policy only if we first go beyond economics, failing which we lack the perspective from which it can be judged.

The ecological movement therefore need not get involved in the quarrel raging between the enemy brothers Market and Plan: both of their clans do nothing but count and accumulate for the purposes of production and economic–political power. Ecology's business is, on the contrary, to prevent political economy from invading everything – starting with the mind, which can be done by sorting out what belongs to necessity and freedom in this realm. Thus, as it will do to corporate profits, it will have to put an end to the State's drastic levy on the GDP by all kinds of means, such as inflation and taxation, by transferring to nature and spontaneous consumption what is being sacrificed to industrial and military investments. It will have to wrest from the imperialism of the market all the previously free or 'priceless' goods, to give them back to nature or to private life. *Écolos* must deal with it all the more, as one of the consequences of their action is that it brings about an accounting and a pricing of what was never before put into figures and assigned prices (e.g. stream water and landscapes). Taking into account all the costs of development must not end up creating a totalitarian market controlled by the State. Its 'containment' might, on the contrary, begin in the area of art – this high-end grocery – and in that of leisure, both of which do not fall within the ambit of basic necessities. And driving back the market implies doing the same with advertising, which puts everything up for sale and artificially generates demand.

Self-management and ecological self-sufficiency

The breaking up of the economy into small self-managed, self-sufficient units for the purposes of freedom and diversity would also contribute to relieving the pressure of a planet-size market while alleviating one of the plagues of industrial society: transportation, one of the main causes of the destruction of nature and societies. There again, the *écolo* sees himself forced to break with one of the nineteenth-century myths common to socialists and liberals: that of a worldwide division of production – another correct idea gone wrong by becoming the only one. Generalized international competition, simultaneously generating wealth, crises and wars, ought to be replaced by another market where countries and peoples (perhaps in the guise of *potlatch*) would exchange, aside from a few high-consumption products, the incomparable surpluses of highly diversified local economies. Thus would disappear the market for cheese, the conditioned protein which, from Hong Kong to Ambert,[12] tends to plasticize everything, to be replaced by fourme, cabrales,[13] chabichou,[14] and so on, and maybe someday the Bandiagara chèvreton.[15]

But entering the path of self-determination and self-management means going against the grain of our time's truths and practices. Thus it is hard to see how the independence of individuals, families and communities could be ensured without a rehabilitation of property. The question is which property, not to confuse it with the one that deprives other people of it but, on the contrary, to call by

---

[12]Translator's note: A town in Auvergne known for the fourme d'Ambert blue cheese, going back to Roman times.
[13]Translator's note: A Spanish blue cheese from the province of Asturias.
[14]Translator's note: A goat cheese from Poitou.
[15]Translator's note: Charbonneau seems to be making up a Malian equivalent of this typical cheese of southern France's Massif Central.

that name small and mid-size property, be it personal or communal, which is physically and spiritually possessed by its owner like the clothes on his back. When the latter no longer inhabits it and sees in it only a capital, this so-called property should escheat and be given back to its true owner, who is its tenant. Among other things, one would thus contribute to the maintenance and restoration of rural real estate.

The security and stability without which freedom cannot take any other risks entail the end of inflation, whereas individuals and communities could build equity that would help them guarantee their security on their own and become independent of capitalists and the State. Savings would replace credit that turns the consumer into the dupe and the serf of a system for which inflation is oxygen, something which would, by the same token, limit superfluous consumption and production. If in addition, instead of constantly rising, tax levies were gradually restored to firms and persons, their powers would increase in proportion to their income. Instead of depending on government subsidies, individuals and local communities could dedicate these surplus funds to social creativity. While they are now the administration's clients and beggars, they would then become used again to the responsibilities and risks of a free man and a free society.

This reclaiming of the economy by nature and freedom – the 'containment'[16] of value creation,[17] autarky and self-management (of the polyculture farm in particular) would contribute to solving the problem of employment, insoluble in a system that duly creates ever more unemployed workers in order to have a pretext for multiplying automated businesses. Once work no longer is, like leisure, a branch apart from life, men and women could devote themselves to activities that have no more to do with water skiing than with the factory or the office. They would no longer be workers or unemployed, but painters, gardeners, househusbands or housewives, fathers or mothers. They would thus leave employment statistics.

---

[16]Translator's note: Charbonneau spells this English word as '*containement*' to turn it into a French one.
[17]Translator's note: Charbonneau writes '*mise en valeur*' – see note 11 above about this term's undertones for him.

As the opposite of a political economy that socializes the individual and brings society under State control, political ecology personalizes society and socializes the State, for it is neither capitalist nor socialist: it is elsewhere. But it will no more escape certain economic necessities than a minimum of State direction of the economy. Only it will not be the same one. And since it will deal with the development of science and technology that determines that of the economy, this direction will actually be one. This State intervention, cautious and wary because it will now be truly practised with its withering in view, is going to be especially necessary during the transition period: like trees, societies and people need a support before they grow back.

Such a radical conversion is not going to be easy: how much restraint and lack of prejudice will be needed in order to succeed! How do we prevent, for instance, the return to self-management from frittering away in a dust of petty fiefdoms, calling in turn for the Prince's intervention? How do we slow down an economy in freefall without blowing up the engine or driving off the road? How do we reincorporate into the earthly ecosystem without tragedy so-called urban monsters of over ten million inhabitants? Beware of a super Cambodia! Before they find themselves unemployed, professionals of the economy and administration will have the opportunity to exercise their talents. Despite a glaring emergency, it is only very gradually that it will be possible to perform such an about turn, after many conflicts and compromises with large interests and the public's habits (let us only think of the car) and with mythologies, such as ideological and nationalist passions. To take on such an adversary with our eyes open, hope is but a feeble help; it will take faith in the meaning and necessity of that enterprise. But the choice is between the latter and nothingness.

The runaway machine must be stopped before it crashes into a wall. This means we have to wake up the driver if he is not dead already. This will not be done overnight like those short-lived revolutions that purport to change everything in a flash. It matters little if this one lasts a hundred or a thousand years, it is its direction that counts. Lowering the growth rate to restore equilibrium, as C. Amery writes, is something that can begin even today in the office of the industrialist, the scientist or the revolutionary thinker, at the garage, at the table or in the garden of the average

Frenchman, in the community field or at the local union meeting. The road without end begins at our feet.

Agricultural policy and the dis-organization of leisure

Politics being the art of managing together, since there can be no question of providing, even condensed into slogans, a program for every ministry, only two key sectors will be discussed here: those of agriculture and leisure, where ecological future planning is either inadequate or nonexistent. For the reasons already laid out, it is important to put an end to the transformation of agriculture and its extension in cooking into agri(?)business. Failing this, since agricultural space – not counting the coastal fishing zone – represents most of the space, the human universe would be transformed into a total suburb peopled in varying densities with suburbanites of higher or lower standing. As it would control food and cooking along with the countryside, this system would no longer leave any part of space-time to human nature, freedom and diversity. If hell is a hermetically sealed world, this one deserves the name.

It is the whole of agricultural space and policy, the rule and not the exception, that must be transformed if we want to modify this or that detail. Rather than islands of 'organic' agriculture and food, it is all agrochemical production that must be gradually restored to agri-culture, even if this return to the land is only partial, mechanics and chemistry still having their uses when their harmlessness is obvious. Once again, it is not a matter of compensating the industrialization of everything with a few ghettos of pure nature, but of driving back as much as possible the totalitarian enterprise out of the most everyday portions of space-time and life by locking it up in rigorously limited national industrial parks. The goal is not an Eden whose apples could be consumed in all innocence as perfectly natural, but an earth where one could at least eat them with their skin.

The return to agriculture – sanctioned by the prohibition of calling 'agricultural' any agrochemical product – would allow massive and immediate savings in mineral fuels, thanks to a use of solar energy that the revolution – that one really green – of the nineteenth century already developed, while it remains to be done in other areas. The rebirth or improvement of genuinely agricultural techniques would by the same token benefit a Third World where the industrialization of agriculture threatens to precipitate the ruin of a particularly fragile pedological and social fabric: you do not jump without transition from the year one thousand to the year two thousand. Agriculture proper would provide the mass of consumers at once with food of varied qualities (not appearances) and inhabited landscapes where it feels good to live and to wander because their beauty changes every moment; finally, it would maintain the cultural richness of Europe by preserving local settlement and an employment base that an automated industry and tertiary sector are powerless to guarantee.

Once the direction has been chosen, the reforms will be self-evident. Savings in energy, environmental cleanup, conservation of nature, air and especially water,[18] the defence of landscapes or the social fabric, unemployment benefits, physical and mental hygiene justify a major transfer of revenue in favour of all farmers – and not just those in the mountains. The exception would be a few highly mechanized farms of the great plains ensuring survival stocks. Of course, the principle of property-as-possession would reserve forest and farmland for those who maintain them as they run them. Like everything that changes life for the better, nothing has to prevent this re-evolution from beginning right now in the lowliest guises. Why should fishers and hunters not take part in the financing of an activity that preserves fish and game by contributing so much per hectare of organic agriculture or agriculture proper? Why should Babylon not finance the garden without which it would choke? The tourists or secondary residents who damage their environment or allow it to deteriorate should pay for this desertion in money or through escheat. Conversely, someone who conserves it should get so much per hectare in the form of tax relief for fields and woods,

---

[18] A vast agricultural space is necessary for the health of water tables. We must therefore view it as a water producer. If we stick to ordinary economic profitability, a high mountain pasture left to mountain breeding brings in less than a ski station. But it produces spring water. At the rate of one meter of rain per year, this adds up to an output of 10,000 m^3 per hectare that the station would soil. At the price of pure water sold in bottles or purified at a great cost, mountain breeding is by far the most profitable.

per are for gardens, per square metre for buildings, which would lead to a positive tax for communities and families of farmers whose work makes landscapes.

Since one reform leads to others, this one assumes an education, in particular that of taste, whose absence explains the passivity with which the public has accepted the disappearance of chicken, bread and fruits, among other foods. This could begin in canteens where children would begin to tell apart a *brugnon* from a fallacious 'nectarine'.[19] To give back a taste for water, which has now disappeared even from mountain resorts, we would reinstal in towns Wallace fountains[20] whose water would come exclusively from the last water tables preserved by the return to agri-culture. And so on.

Such a policy is particularly needed in an urbanized and overpopulated Europe that would perish without a vast green lung and whose cultural richness is rooted in the diversity of its countrysides. As much as of national development plans, these have been victims of the Mansholt Plan, however unpleasant this might sound to this ecologist who 'has not signed a lease with truth'. Instead of a Europe of the Golden Delicious apple and the factory calf, why not have that of peasants and their fruits? Why couldn't we align European agriculture with the laws of the country where nature and agriculture are the best protected, instead of with the most 'competitive' agrochemical sector? This would be a first step in the right direction.

The countryside is also the business of the city which, it bears repeating, is located in a territory whose sap it extracts and where citizens can get out. This leads us to a different politics of space, one that would let it be for them both a natural space and a space of freedom. Since any living being needs a minimum of territory, couldn't this fact be registered by law, both nationally and internationally? Why not so many ares per family, where everyone could cultivate his garden? This would have, among other advantages, that of prohibiting the powers that be and the

---

[19] Translator's note: In French, a *brugnon* differs from a nectarine in that its pit adheres to the fruit's flesh, whereas both are called nectarine in English and other languages.
[20] Translator's note: Ornamental cast-iron public drinking fountains first installed on Paris streets in the late nineteenth century and emulated since then in smaller samples in many other cities in France and abroad.

promoters from indefinitely piling up people on a vertical axis, since they would automatically have to make available all the more space for private or public gardens: many cities provide the example of a stone labyrinth whose doors open onto a secret countryside surrounded by houses.[21]

A versatile and natural activity, agriculture – not the one that consists in running around in circles all day long with one's ass on an engine – solves the problems of a society in thrall to the work–leisure dichotomy. A peasant population living in its countryside would no doubt have less reason to go and find nature again in Nuku Hiva[22]: travel would no longer be the false way out that society opens at a set time. Ecology encounters mass – hence organized – leisure at every turn of its defence action. As the manufacturer of a superfluous that can be indefinitely increased, the leisure industry is one of development's engines and, since nature is its object, the first cause of its devastation. Only war can waste even more energy and space. But this standardized leisure, concentrated qua organized, has no reason for being, but for the profits of tour operators.[23] For its justification is to provide to all what it annihilates: nature and freedom. What it sells is in the final analysis the cement wall at the foot of which people and cars pile up more than in Manhattan, a life even more price-tagged and Taylorized than at the assembly line where the skilled worker at least tries to slack off. The worst material and especially spiritual side of neo-industrial society is not work but leisure as heavy industry. Within it, pollution, crowding, the lie of a class struggle where the obsequiousness of the workers is only matched by their hatred of tourists, the cartoonish display of bourgeois ugliness and conformity extended to the people, all rot in broad daylight. It is not Billancourt, but Saint-Tropez in August that could make one despair of mankind.[24]

---

[21] Bourges, among others.
[22] Translator's note: The largest of the Marquesas Islands in French Polynesia.
[23] Translator's note: In English in the original.
[24] Translator's note: Bernard Charbonneau is playing here on Jean-Paul Sartre's famous dictum dismissing any discussion of the horrors of Soviet totalitarianism, typical of the post-war attitude of most French intellectuals of their generation: '*Il ne faut pas désespérer Billancourt*,' 'We must not drive Billancourt to despair.' It alludes to the Renault car factories in this Paris suburb as shorthand for the working class and its trust in Communist revolution as the promise of the future and the meaning of life.

It is therefore urgent that we dis-organize leisure to give it back to itself, that is, to nature and freedom. Otherwise, within ten years, it will have soiled the last places of our earthly paradise. We must wrest leisure from the claws of its sundry organizers, dealers and developers, to give it back to human spontaneity. Since its realm is not the necessary but the superfluous – except for those who earn their bread or their caviar from it, leisure is not indispensable to a vital minimum in the way that work is. It may thus be excluded from the market and from economic organization without serious drawbacks. Any tourism propaganda or organization would be considered as an offence against freedom and nature, while leisure would be left to the imagination or to risk: you would be free to get killed in the mountains or at sea and not be guaranteed a rescue. And as far as possible, any kind of cement or motor would be excluded from this sector. The credits dedicated to tourism development, in particular to the highway kraken, would be transferred elsewhere. As for tourist areas, instead of pimping out their beauty and their originality, they would be allowed to revert to being places like any other, living from agriculture or industry.

The dis-organization of leisure would make it possible, among other things, to get out of the dead end of a tourism development that here and there turns into a pandemonium. For if 'people' found at home what they vainly seek ever further afield, a great many of them would not give in to the summer panic. Besides, since organization concentrates in time and space, a de-standardization would lead to a diversification of leisure that would disperse the masses. Instead of concentrating in Saint-Tropez just to be there with and in front of other people, one would be a gardener in Les Andelys,[25] a carpenter in Bourganeuf,[26] a violinist in Montmartre or a fisher at the Vert Galant[27] in a Seine of actual water, and so on. Thus would emerge in leisure itself a different kind of work, another way of life, whose quality products, such as those of the farmyard, the kitchen garden and the workshop, could be given or exchanged; the greater part of survival production being the business of a civilian service designed to keep automated industry under control. And to the degree that, little by little, leisure games

---

[25] Translator's note: A town on the river Seine in Normandy.
[26] Translator's note: A town on the river Creuse in central France.
[27] Translator's note: A green space in the centre of Paris.

would be reintegrated into more human and more varied work, the distinction would fade. But we must first make Mr Trigano harmless.[28]

Many other programs would need to be provided to the future ministries. But the question is more how to break free of them. Thus, instead of proposing a reform of education that is also imposed from the top without success, we would do better to try and figure out how to gradually give it back to persons, to families or to communities of teachers and those they teach, which would take on the responsibility of empirically opening the way for genuine change. The Beautiful and the Good cannot be defined in advance by any policy; its task is merely to modestly realize some of their conditions. It may be said that the policy that would establish an equilibrium, deconcentrating and demassifying our society, would realize the conditions of freedom and equality, and not, under their label, those of servitude and inequality. As for foreign policy, this distant suburb of inner life, since it belongs to the realm of pious wishes at this stage of the *écolo* movement, it is not here that the reader will find out how universal peace can be ensured by discussing with Messrs Brezhnev, Carter and their ilk. Ecology will have plenty of time to do this once it is installed in Giscard's armchair. Let it remember that the seed is not the tree.

---

[28]Translator's note: Gilbert Trigano (1920–2001) went from Communist journalism to the development of Club Med by way of the family tent-making business that first supplied the equipment for its tourist villages in 1950.

Envoi

Author and reader have now come to the end of their task: the book's writing and reading. By way of farewell or goodbye, the former may now talk to the latter in person, that is, by speaking in the first person. I hope that my *écolo* companion – or my companion, period – will excuse me for having gotten personal; there is no other way to be objective, breaking with the commonplaces that pollute the air we breathe. May he forgive me for having perhaps discouraged him by pointing out the towering height and the difficulties of the task at hand: one does not set out for the Matterhorn wearing sandals. And I am not in charge of dealing out dreams to youngsters and sleeping pills to the elderly. The question of the possible and the impossible comes only after that of the true and the false – yours, I hope. It may be that ecology is but one more social mystification; it is even likely, the realistic mind will think. But if that were to be the case, the master of nature or of creation would simply have shown that he was incapable of mastering his own nature.

The most modest ecological enterprise demands no less of man than that he vanquish his own nature, which is social. I am the first to know it. But this is not the issue. Where are we headed on these oh-so-comfortable tracks? To the unnameable (although it is called nothingness). From the first day, which starts again for everyone each dawn, man has been condemned to overcome himself, like an atomic rocket programmed to self-destruct if it does not reach its goal. In this case, it is not the destruction of the earth, but the creation of freedom.

INDEX

advertising 32, 87, 101, 135, 168, 180–1, 183
Africa 20
agriculture xxii, 26, 61, 107, 179
  agribusiness 6, 16, 26, 107–10
  industrialization of 107, 153, 187
  organic 27, 99, 107–8, 153
  space and policy 186–91, 187 n.18
agronomics 11 n.20, 116
Algerian War 9, 12
Alps 5, 169
Alsace 22
Amazon forest 82
Amery, Carl 65, 72, 74, 185
Amoco Cadiz oil spill 26
amor fati 4, 58
anarchism 7 n.12, 22, 30, 37, 40, 71, 79, 80, 88, 90, 98, 110, 117, 123, 131
Anglo-Saxons xiii, 5, 21, 33
Anthropocene ix, x, xxxiii n.8
anti-Americanism 6, 110
anti-authoritarian xi
anti-Christian 36, 73, 80
anti-colonialism xxiii
anti-hunt 40, 45
anti-militarism 45
anti-modernism 81
antimorals 37, 157
anti-scientism 102
anti-Sovietism 110
anti-statism 45

apocalypse 99
Aquitaine 24, 121
Aquitaine coast 11, 89, 169
Arbor Day 24
Aristotle 69
army xvii, 8, 23, 29, 32, 37, 48, 90, 114–15, 133, 155 n.1
  see also military
Aron, Raymond 10
Asia 20, 81
Association pour la protection contre les rayons ionisants (APRI) 22
atheism 72, 147
Atlantic Ocean 15
atomic age 6
Audubon, James 16
authorities xxxiv, 23, 31, 82
authority xxxv, 18, 23, 37, 63 n.2, 72, 82, 84, 88, 91, 102, 110, 116–18, 131–2, 144–5, 158
  vs. power 167–8, 178
automobile xxxiv, 108 *see also* car
Auvergne 183

Baader-Meinhof Gang 90
Babel 70, 165
Babeuf, Gracchus 7, 35
Babylon xxiii, xxxii n.5, xxxiii n.9, 4 n.2, 5 n.3, 56, 65 n.1, 70, 187
Bacon, Roger 72
Baikal, Lake 20
Balzac, Honoré de xxxii

Barjavel, René 6
Barthes, Roland 173 n.3
Basques xxxvii, 28, 32, 37, 82, 147 n.1
Béarn xxxii–xxxiii, 39, 167 n.13
Belgian Revolution 124
Belgium 169
Belle Époque 6
Bercé forest 3
Berry, Wendell xvii
Biafra 147
Bible, the 67–9, 76
biology 13, 16, 80, 82, 96, 102, 161
Blut und Boden 8–9
Boiteux, Marcel 150
Bonnefous, Édouard, 11
Book of Job 69
Bordeaux xxii, xxxiii n.8, 88 n.6, 150 n.3
Bordeaux School xiv n.6, 12 n.24
bourgeois, bourgeoisie xvi, xxxvi, 5, 16, 19–20, 23, 34, 45, 86, 88 n.6, 95, 100, 105, 109 n.17, 114, 117, 121, 143, 145, 147, 150, 142, 156–7, 160, 189
  intellectual bourgeoisie xxxvi, 45, 160
Bourges 189 n.21
Brazil 36
Bretons xxxvii, 31, 82
Brezhnev, Leonid 191
Brocken 8
Brussels 124
Bucharest United Nations World Population Conference (1974), 20
Buddhism 31, 68
Bugey demonstration 31
bureaucracy xiv, xxv, 119
Burgundy 100

Burke, Edmund 7, 79
Burnham, James 11
Burundi 110
Byzantium 172

Cambodia 129, 185
canton 6 n.6, 90 n.9, 100, 107
capitalism 9, 37, 57, 67, 89, 91, 104, 106–7, 109–10, 119–20, 129, 133, 146, 158, 169, 209
car xxxiv, 9, 12, 21, 56, 61, 99–101, 108, 121, 151–2, 185, 189 *see also* automobile
Carson, Rachel 16
Carter, Jimmy 21, 191
catastrophism 99
Catherine the Great 171
Catholics 5, 11, 33–6, 72, 82
centralization 104, 128, 134–5, 161, 163, 175, 182
centre (of centralization) 114–15, 129–30, 133, 153, 168, 169, 189–90
Centre (on political spectrum) 30, 150
centre-right 6 n.8, 134
CFDT (French Democratic Confederation of Labour) 35, 134, 177–8
Chaban-Delmas, Jacques 23
Chamber of Industry and Commerce of the Federal Republic of Germany 72
chaos xxv–xxviii, 56, 62, 97, 144, 149
Charbonneau's books:
  Dimanche et lundi 105 n.9
  L'État xxiii, xxxiii
  Le Feu vert xxiv, xxvii
  Finis Terrae xxiv, 101 n.3
  L'Hommauto xxxiv n.10
  Il court, il court, le fric 107 n.12

INDEX

Le Jardin de Babylone xxxiii
  n.6, xxxiii n.9, 5, 4 n.2, 5
  n.3, 65 n.1
Je fus. Essai sur la liberté xxiii,
  xxiv, 155 n.1
*Nous sommes des
  révolutionnaires malgré nous.
  Textes pionniers de l'écologie
  politique* (with Jacques
  Ellul) 4 n.2
Nuit et jour. Science et Culture
  101 n.4, 181 n.11
La propriété, c'est l'envol
  107 n.12
*Le Système et le Chaos. Critique
  du développement exponentiel*
  xxiv, xxxiv n.10, 62 n.1
  translation of x, xviii–xix,
  xxviii–xxix
Tristes campagnes xxiii, xxxii
  n.6
Un festin pour Tantale 109 n.16
Charlie Hebdo 23, 25 n.7, 29
  n.2, 30, 37
cheese 109, 183
Chevalley, Claude 31
Chile 40
China 9, 36, 129
Chirac, Jacques 118
Christianity 4, 11, 21, 26 n.9, 36,
  58, 65–9, 71–6, 81, 89,
  109–10, 147, 155 n.1
  Christian idealism 73–4
  condemnation and glorification
    of nature 69–71
  on creation 65–9
  ecological movement and 72–6
  rigorous tendencies of 72
  rupture between Adam and
    nature 67
  temptations 71
cities, city xxxiii, 3–5, 8, 20–2,
  25, 31, 61, 38, 35, 56–7, 59,
  61, 69–70, 72, 85, 107, 122,
  132, 185, 188–9
class struggle 91, 189
Club Med 101, 191
Club of Rome 11, 24, 100, 116, 143
Coleridge, Samuel Taylor xiii
collective behaviour xxii
Columbus, Christopher 18
Comédie-Française 124
commonplaces 74, 173 n.3, 193
communication 69, 159–60,
  163–6, 168
communism xxiv, 9, 11, 35, 38,
  41 n.20, 114, 129, 143, 158,
  182, 189 n.24, 191 n.28
Communist Party 10–11, 23,
  31, 34, 36, 129, 135, 145,
  173, 177
computers xxv, xxxviii, 24, 46, 62,
  70, 119, 147, 153
Concorde 15, 26, 74
Conseil pour la protection de la
  nature 17
conservation 24, 34, 119, 187
conservative or revolutionary
  32, 126
conservative revolution, 149, 151
conservative right 92
consumer society 106
cooking 107–8, 186
Côte d'Azur 180
Council of Europe 23
counter-society 131, 141–2
countryside xxxii, 4, 11, 20, 24,
  33–5, 40, 56–7, 85, 89, 107,
  114, 118, 132, 153, 166, 169,
  186, 188–9
crafts, craftsmen 3 n.1, 7, 32, 174–5
Creuse River 190
culture xix, xxviii, xxxvii, 10,
  24–5, 53, 57, 59, 74, 80,
  83, 85–6, 109, 122–4, 142,
  147, 149

INDEX

dams 10
DATAR 116
DDA 31
death 4, 9–10, 17, 19, 23, 29, 49–53, 58, 64, 67–8, 71, 76, 80–1, 96, 99, 109 n.17, 144
death of God 92, 126
death of man 126
Debré, Robert 181
decentralization 103, 191
defence committees 22, 27, 39, 99, 130, 151, 161–2, 169
De Gaulle, Charles 29 n.2, 123 n.9, 179 n.7
degrowth 180
demographics 20, 110, 152, 181
  see also population
depoliticization 38, 114, 128–33, 137
Deuteronomy 70
developed countries 15–16, 19, 123, 181
  feeling for nature in 4
development 22–5, 31 n.8, 34–6, 45–6, 51, 61, 65, 72, 74, 89, 95–101, 104–5, 110, 116 n.5, 118–21, 126, 128–30, 132 n.5, 135, 143, 145, 149, 150 n.3, 152–3, 160, 174, 177–82, 185, 188–90
dietary science 109, 158
disincarnation, discarnate 46, 48, 66
Domenach, Jean-Marie 25
Dorst, Jean 16
Duboin, Jacques 106
Duhamel, Georges 6
Dumont, René 11, 26, 27, 109, 117
Duverger, Maurice 11, 125 n.3

Eastern mysticism 33
Eastern spiritualities of detachment 7
Eastern wisdom 87
Ecclesiastes 69
ecodevelopment 179
ecofascism 82–3
ecological enterprise 151
ecological leftism 35, 86–90, 149
ecological meeting 161–6
  in form of a colloquium 164–5
ecological movement xxxvi, 7–8, 12, 15–16, 28–9, 51, 95, 129, 131, 133, 145–6, 149–50, 152, 156, 176–7
  absence of reflection in 100–2
  antinuclear struggle 28, 45, 62, 98–9, 105, 107
  biology and genetic engineering, case of 102
  Christianity and 65–9
  communities and 38–41
  counter-society and 141–2
  critique of anarchistic and non-violent strand of 88–92
  critique of 'development' 97–100
  emergence of a naturist revolution 3–8
  in France 22–8
  function of 146–7
  left-wing ecologist vs. right-wing ecologist 34–8
  libertarian left, tone given by 86–8
  as a neurotic rebellion 32
  prehistory of 3–8
  principles 92
  recycling and 117
  shortcomings in economic and social matters 105–11
  weak points of 97–105
ecological order xix, 157, 167–9, 178
ecological party 135, 167, 178
ecological politics 134–5, 145, 153, 162, 171–91

INDEX

ecological problem 23, 26, 169
ecological reaction xxiii
ecological revolution 137–8,
    142–4, 161–3, 174
ecological sects 33
ecological self-sufficiency 183–6
ecological society 161–6
ecological union 147
ecological u-topia 148–54
ecology 17–18, 25
    approach to mediation between
        opposites 146–8
    core issues of 143
    economic policy and 179–83
    label of 95
    power and 171–6
    as the science of natural
        equilibriums and
        harmony 157
    without a doctrine 95–7
écolo movement 120, 124, 137,
    149, 153, 158, 160, 168,
    172–5, 177–8, 182–3,
    191, 193
economic development policy
    100, 179–83
economics 85, 89, 95, 96, 100,
    105–6, 110, 115–16, 118,
    123, 126, 128–9, 133, 141,
    143, 145, 150, 152, 158, 175,
    179–83, 185, 197, 190
ECOROPA 12 n.24
EDF *see* Électricité de France
elections 40, 114, 127, 134–6
electoral democracy 128
Électricité de France (EDF) 36,
    135, 180 n.9
electronics 15, 84, 86, 119, 121
Ellul, Jacques ix, xi, xiv, xxi–xxiii,
    4 n.2, 7 n.11, 11, 26 n.9,
    173 n.3
Emerson, Ralph Waldo xiii
empire 48, 161
England xiii, 8–9

English Channel 169
environment x, xvii, xviii, xxxiii
    n.8, xxxv, 8, 10 n.5, 13,
    17–19, 21, 24–5, 35, 39, 45,
    47, 55–7, 59, 84, 100,
    116–17, 119, 126, 130, 135,
    151, 156, 161, 172, 175,
    182, 187
environmentalism ix–x, xxvii
environmental philosophy x, xii
environmental thought xiii, xvii
equality 7, 15, 30, 35, 73, 86–7,
    128, 156, 162, 174, 191
Eskimos 82
Esprit xiv, 25, 166 n.11
Esso Standard 30
ethics 82, 86, 157–8
ethnology/ethnologists 16, 19, 25,
    81–3, 142
ethology 82
Eurasia 20, 68
Europe 9, 11, 18, 20, 28, 39, 56,
    85, 107, 110, 135, 143, 163,
    169, 178, 187–8
European federation 12 n.24, 169
Europe Écologie 178
L'Expansion xxxi
extremism 31, 59, 114, 150

Fabian Club 167
Fallières, Armand 31
family 23, 37–8, 49, 71, 79,
    88–9, 107–8, 111, 150, 156,
    161, 173
farmers, farming 22, 27 n.12,
    107, 109, 116 n.3, 123, 184,
    187–8, 190
    industrial xxv
    organic xxxv
fascism xiv, xxiii–xxiv
fashion and fashionistas xvii,
    xxvii, xxxvii, 6, 20–3, 26–7,
    30, 47–8, 58, 73, 101,
    115–18, 148, 155 n.1

INDEX

Faure, Edgar 23
Fédération française des sociétés de protection de la nature 22
Feuerbach, Ludwig 108 n.15
FIAT 11, 117 n.6
Fifth vs. Third Republic 61, 130, 179 n.7
Figaro, Le 12
First World War xxi–xxii, 6, 8, 91
fishers 39, 122
fixation abscess 8, 133
Flaubert, Gustave xxxvii
food xviii, 32, 40, 53, 66, 70, 107, 108–9, 154, 158, 166–7
Fourastié, Jean 26, 84
Fournier, Pierre 23, 25 n.7, 99
France xxi, xxxiii, 6, 9–13, 21–5, 26 n.10, 27, 38, 39 n.19, 56, 107–8, 110, 116, 119, 127–8, 130, 147, 162, 173 n.3, 188 n.20
France, Anatole 6
La France défigurée 25
Franciscan mysticism 72
freedom xi–xvi, xxii–xxiv, xxvi–xxix, xxxiv–xxxvi, 7, 36–7, 48–50, 63, 80, 146, 156, 178
  as ally of nature 60–3
  contradiction between nature and 50–4, 59–60
  definition in ecological struggle 45–6
  and equality xiii, 7, 15, 30, 87, 128, 174, 191
  and nature xxi, xxviii, 9, 92–3, 97, 101, 104–5, 123, 129, 142, 146–8, 151, 178, 184, 189–90
  price of 64
  pure 66, 111, 162
freemasonry 167
French environmental movement xxiii

French political ecology xxi
French Revolution 49, 88, 114, 142, 162, 175
Freud, Sigmund 30, 81
Friends of Man 131
Friends of the Earth 21, 25, 76
future xxvii, 6, 23, 35, 75, 80, 83, 117, 120–1, 126, 128–9, 174, 179–80

Garreau, Gérard 109
Gascon personalism xi, xxi, xxiii, xxiv
Gascony 147
Gauguin, Paul 5
Gaullism 34 n.10, 134 n.6
General Motors 21
Genesis 66, 76, 93
Geneva 4, 73
gentle technology 100, 103–4
  see also soft technology
geography xxv, 29, 121, 151
geology x, xxii, xiii
Germany xiii, xvii, 8, 46, 72, 97, 123, 130, 169
Giono, Jean 5
Girard, René 110
Giscard d'Estaing, Valéry 6, 36 n.13, 191
Gorz, André 180 n.9
La Grande Bouffe (Marco Ferreri film) 109, 158
Great Moulting (*Grande Mue*) xxi–xxiii, xxxiii, xxxiv–xxxv, 9, 13
Greeks, ancient 46, 66, 76, 81, 157
green lie 118
Green Revolution 11, 186
Greens xiii, 173 n.3, 178 n.6
green space xxviii, 45, 57, 83, 190
Grothendieck, Alexandre 31
Groupe du Chêne 167 n.13

growth xiv, xxxiv, 8, 10, 16–17, 20, 24, 26 n.10, 60, 84, 99–100, 119, 143, 179–80, 185
Gueule ouverte, La 25 n.7, 40

Haeckel, Ernst 16–17
Hainard, Robert 80, 85
Hamburg market 33
Hara-Kiri 29
Haydn, Joseph 66
H bomb 9, 74, 147
Heidegger, Martin 181 n.11
Heim, Roger 16, 117
Hervé, Alain 30 n.4
highways xxxi, 16, 24, 26, 35, 39, 56, 61, 90, 130, 190
Hinduism 68
hippy movement 21, 31
Hiroshima xxxiii n.8, 6, 9, 47
history xv–xvii, xxi–xxii, xxv, xxix, xxxii–xxxiv, xxxvi, 3, 9, 11–13, 30, 35–6, 55, 60, 68–9, 71, 74–5, 81, 96, 99, 131, 137, 148, 151, 155 n.1
Hitler, Adolf 10, 137, 142
homosexuals 37, 178
Hong Kong 183
human ecology xxxvii
human rights 155 n.1
human sciences 102–3, 144
L'Humanité 12
hunting 17, 40, 92, 122, 175, 187
Huxley, Aldous 7, 86
hygienism 8

IGAME 24
Illich, Ivan 7, 11, 27, 35, 110, 117
incarnation 45, 48, 59, 69, 74, 76, 93, 156
Indians 19–21, 71, 82, 123
individual xiii, xxii, xxiv–xxvi, xxxvii, 4, 7, 12, 13, 17, 19, 27, 45–50, 52–3, 59–60, 62–3, 68, 72, 80–2, 85, 88, 90–3, 103, 107, 118, 120, 124, 127, 131, 143, 147–8, 155–7, 159–62, 165–6, 171–3, 183–5
individualism xi, xiii, 5, 45, 60, 73, 156
individuality xxii
individualization 59, 63
Indochina War 9
industrial development xxxiii, xxxviii, 8–9, 150
industrialization x, xiii, xvi, 34, 107, 109, 153, 186–7
of agriculture 107, 153, 187
industrial society xi, 3, 5, 23, 26, 32, 63, 103, 105, 108, 113–14, 128–9, 135, 151, 164, 183, 189
opposition to 18, 20, 101
second 8, 15
information xiii, xxxvi–xxxvii, 25, 98, 117, 136, 164–5, 168
information technology (IT) 104
INRA 116, 118
intellectual bourgeoisie xxxvi, 45, 160
intellectuals, intelligentsia xxi, xxiii–xxiv, xxxvi–xxxvii, 5–7, 9–10, 13, 19, 26 nn.9–10, 33, 45, 80–1, 98, 115–16, 118, 127, 129, 131, 134, 144, 156, 160, 168, 173 n.3, 189 n.24
irrationalism 80–1
Islam 127, 147

James, William xiii
Jansenism 72
Jaulin, Robert 19
Jehovah's Witnesses 131
Jesus Christ 67–8, 71–3
Jeunes et Nature 25
Jews 65–6, 147

INDEX

Jones, Jim 131
Journalistes et Écrivains pour la protection de l'environnement 25
Judaeo-Christian tradition 69, 74
Jugendbewegung 8

Keynes, John Maynard xxxiii, 10, 34
Klein, Naomi xv
Krupp xvii, 46

labour xi, xxiv, 37–8, 105, 142
Labour Party 167
La Fontaine, Jean de 4
Lalonde, Brice 28
land consolidation 26, 61, 90, 107, 116
landscape xxv, 12, 15, 17, 24–5, 56–7, 83, 86, 107, 116, 119–20, 151, 182, 187–8
language xxviii–xxix, xxxv–xxxvii, 23, 61, 75, 85–7, 131–2, 159–61
Languedoc 147
Larzac 27, 34, 36, 40, 98
Latin countries 5, 10
Lautréamont 88 n.5
Lawrence, D. H. 80–1
Lefebvre, Mgr Marcel 31
Left/left-wing 7–8, 10, 20, 23, 28, 30–1, 34–6, 38, 74, 79–80, 83, 86–90, 92, 95, 98, 106, 109, 113, 115, 117, 127, 129, 134–5, 148–50, 153, 172–5, 178
  intellectuals xxxvi, 10, 26, 156
  left-wing naturist fundamentalism 79–80
Leibniz, Gottfried Wilhelm 72
leisure xvi, 5, 39, 105, 107, 142, 157, 175, 179, 183–4, 186, 189–90, 190

Lemoniz nuclear plant 28
Lévi-Strauss, Claude 19
Lévy-Bruhl, Lucien 19
liberalism xiii, xiv, xxiv, 10, 21, 23, 73, 128, 135, 147, 182–3
libertarian left 35, 37, 51, 73, 79, 86, 92
liberty xi, xvi, 173
Ligue pour la protection des oiseaux contre les chasseurs 17
Lip factory worker takeover 27, 35 n.11, 36 n.15, 40
love xiii, 16, 37, 48–9, 52, 68–9, 71, 74, 87–8, 152, 162, 171
love of nature 3–4, 45, 47, 58–60, 63, 84, 107, 142
Low Countries 169
Lyon 100
Lyonnaise des Eaux 119

machinery, machines xvii, xxxviii, 7, 9–10, 23, 46, 62, 69, 106, 118, 120, 128, 185
Magellan, Ferdinand 74
Maison de Fleurance 109
Maisons paysannes 22
Maistre, Joseph de 79
Mali 183
Malville nuclear plant 28
Manchester 4
Manhattan 189
Mansholt, Sicco 11, 12, 116
Mansholt Plan 11–12, 116 n.3, 188
Maoism 38
Marcuse, Herbert 30
Marie-Antoinette 5, 123
market x, xxix, 10, 83, 103, 106, 147, 179, 181–3, 190
Marx, Karl 10, 20, 96
Marxism xxxiv, xxxv, 7, 10, 20, 22–3, 30, 37, 40, 89, 98, 162, 181

Massachusets Institute of Technology (MIT) 24
Massé, P. 11, 24
Mauriac, François 88 n.6
May 1968 22–3, 105, 123, 128, 141, 166
McLuhan, Marshall 173 n.3
media xxxvii, 12, 18, 25, 27, 104, 113, 115, 117, 124, 127, 136, 163, 165–6, 168–9
medicine 15, 27, 33, 37, 52, 110
Melville, Herman 5
Midi (southern France) 33
militants 11, 29, 31, 33, 128, 133, 136, 149, 153, 168, 172, 174
military xxxi, xxxiii, xxxviii, 8–9, 17, 23, 29, 31–2, 36–7, 39, 45–6, 48, 83, 90, 72, 101, 111, 127–8, 130, 135, 141, 145, 182 *see also* army
Mill, John Stuart xiii
Mission interministérielle pour l'aménagement de la côte aquitaine (MIACA) 11 n.23, 89
monarchy 127–8, 175
monastic orders 167
Le Monde 11 n.19, 63 n.2, 130 n.4, 163–4
money 21, 70, 106, 119, 131–2, 136, 178, 181, 187
Monnet Plan 11
Monod, Théodore 19
Montaigne, Michel de xxxii
Montesquieu, Charles-Louis de Secondat, baron de La Brède et de 167
morals xiv, 8, 21–2, 31, 34–5, 67–8, 86, 111, 117, 157
Morin, Edgar 11, 25
Mounier, Emmanuel 25 n.9, 166 n.11

Muséum national d'histoire naturelle 16, 18, 36, 85
Mussolini 82

Nader, Ralph 21
natalism 181
nationalism xxiii, xxvi, 20, 35, 37, 62, 110, 173, 185
national parks xxviii, 5, 17, 20, 83, 100, 118, 122, 124
nation-state 122, 135, 147, 181
natural ecosystems 18
naturalists xxviii, 16–17, 19, 80, 83, 85
nature 4–5, 19–20, 46–8, 146, 178
 as ally of freedom 60–3
 condemnation and glorification in the New Testament 69–70
 connection with man 47–8, 50, 53, 55–7
 contradiction between freedom and 50–4, 59–60
 definition in ecological struggle 45–6
 feeling for xxi, xxiv, 4, 45, 62, 73
 love of 3–4, 45, 58, 142
 as material of landscape 56–7
 organizations to protect 22
 as the Other 58
 personal relationship with 58–9
 protection of xiv, xxvii, 22–5, 47, 86, 106, 118, 122
 return to 8, 81, 111
 romantics and 5
Nature et Progrès 22, 27, 116
naturism 8, 31, 33, 35, 58, 83
naturist fundamentalism 131–2
 animal society as a model 82
 critique of ecologism 84–6
 individual, view of 81
 left-wing 79–80
 as a manager of nature reserves or of national parks 83

peasant, view of 82
returning to nature 81
right-wing 79–80
status of women 82
naturists xxxiv, 5, 8, 31, 33, 35, 40, 55, 58, 84–5
Nazism, national-socialism 35, 80, 82, 173
'neither Right nor Left' 172, 173 n.3
New Testament 4, 67, 69–70
Newton, Isaac 72
Nietzsche, Friedrich 80, 102
non-violence 21, 36, 40, 49, 90–1
Normandy 190
Norway 56
Nouvel Observateur, Le 12, 30 n.4, 180 n.9
nuclear arms 9, 20, 47, 60, 74, 76, 85, 91, 99, 108, 110, 129, 144–5, 147, 153, 193
nuclear energy 9, 27, 40, 61, 99, 130, 150, 152–3
nuclear plants 21, 28
nudism 8
Nuku Hiva 189

Office national des forêts (ONF) 108
oil 15, 17, 26, 39, 99, 119, 121, 123, 128, 134–6, 156, 163–5, 167, 169, 171, 175, 177–8, 182, 189–90
oil spills 26, 99
Old Testament 4, 66, 68–9
opportunism 30, 117–18, 131, 173
organization xxvi–xxviii, 21, 40, 61–2, 73, 79, 84–5, 96–7, 103–4
Oriental paganism 66
original sin 67
origins 19, 35, 46, 60, 65, 97, 174
overpopulation 181

pacifism 6, 36, 91, 173
paganism 66, 68, 74, 76, 81
pantheism 58, 81
parapsychology 33, 80
Paris xxxii–xxxiii, 11, 19, 28, 30, 100, 190
parties 10, 11 n.18, 25, 28, 32, 34, 89, 128, 131, 133–5, 148, 173, 177–8
Parti socialiste unifié (PSU) 35–6, 134
Pascal, Blaise 72
peasants, peasantry xvii, xxxiii, 7, 11, 20, 25, 32, 56, 74, 82, 58, 120, 122, 174–5, 188–9
Peccei, Aurelio 111, 117
Pellerin, Pierre 16–17
Pennsylvania Dutch 21
Peoples Temple 131
Périgord 169
person xv, xxiv–xxvii, xxxv, xxxvi, 4 n.2, 12 n.24, 18, 31, 47–8, 52, 57–8, 60, 66, 68, 71, 75, 81, 85, 92, 102–3, 115, 122, 124, 127, 136, 147, 153, 155–66, 168, 173 n.3, 184–5, 193
personalism xi, xiii–xiv, xxi, xxiii–xxiv, xxix, 25 n.9, 150 n.3, 155 n.1, 166 n.11, 167 n.13
Pétain, Philippe 30 n.3, 34, 155 n.1
Peter the Great 9
phalansteries 132
Pilote 29
Pisani, Edgard 116
plan, planning 111, 153, 169, 179–82, 186, 188, 191, 193
police 17, 35, 37, 45–6, 61, 99, 114, 141, 161, 168, 175–6
politicization 10, 38, 128–9, 133–7, 162
politics and political functions 126–9, 134

INDEX

development and 129–30
ecological politics 134–5, 145, 153, 162, 171–91
ecological tactics in 176–8
'everything is political' 36, 89, 133, 135, 172
political commitment 126–7, 129, 135–7
Right–Left game 172–4
pollution 12, 17, 20–1, 24–5, 33, 39, 45–7, 98–100, 107, 119, 123, 143, 161, 169, 175, 180, 189, 193
Polmar plan 26
polyculture 116, 118, 184
Pompidou, Georges 23, 27
population 26 n.10, 143, 181
see also demographics
population explosion 37
Portugal 74
post-Christian 73, 89, 155 n.1, 157
potlatch 183
Pourrat, Henri 6
power xxiv–xxv, 5, 9, 48, 60–2, 68, 70–4, 82, 84–5, 90–1, 96, 99, 102–5, 107, 110–11, 113–14, 120–1, 126–9, 131–7, 142–3, 146, 161, 163–5, 167–8, 171–7, 188
prisons 36, 48, 84, 178
private life 127, 133, 156, 182
private life vs. public life 135, 151, 156–7
progress xxiv, xxv–xxvi, xxxi–xxxii, xxxiv, 3, 5–7, 9–10, 15–16, 18–21, 23, 30, 34–5, 50–2, 57, 62, 68, 72–3, 80, 83, 86, 95, 100–2, 113, 129, 148–9, 151, 173–4, 182
progressives xxxiv, 11, 26 n.9, 35, 40, 55, 69, 80, 84, 152 n.5, 173–4

proletariat 145, 147, 152
propaganda 27, 116 n.5, 126, 128, 134, 178, 190
property 34, 61, 66, 89, 107, 145, 183, 184, 187
property-as-possession, principle of 187
Prophets 70
Protecna 119
Protestantism 4, 20, 23, 72
Protestants 36
Proudhon, Pierre-Joseph 50, 107 n.12
Psalms 70
psychoanalysis 11
psychology xvi, xvii, 18, 85, 105, 119, 151
PTT (French post) 104
publicity 9, 165–6
public relations 160, 164
puritanism 40, 87, 131, 157
Puritans 5, 21, 67, 87
Pyrenees 147 n.1, 169

Quakers 21
Quebec 147

Ramuz, Charles-Ferdinand 6
Rassemblement pour la République (RPR) 34
rationalism 31, 40, 81
rationalization 5, 20, 104, 110
reactionaries 6, 25 n.9, 40, 110, 173
recycling of ecology 113–14
by cultural spectacle 122–4
professional recyclers, role of 117
through depoliticization 129–33
through fashion and fashionistas 115–18
through politicization 133–7
through technostructure 118–22
Red Brigades 90–1

reform 187–8
Reformhäuser 33
reformism xv, 122, 134, 148, 151
regionalism xi, xiv, 6, 28, 37, 45, 61
Reich, Wilhelm 30, 37, 96
religion xxvi–xxvii, 32, 61, 65, 67–9, 71–6, 81–2, 97, 101–3, 110, 126, 131, 142, 144, 146, 148, 172
Renaissance 3 n.1, 4
Renault 151, 189 n.24
revolution 135, 137, 142–3, 148–9, 151, 157, 160–2, 174–5, 185–6
revolutionary because conservative 174
revolutionary conservatism 151
Rhine 169
Rhin-Rhône canal 26–7
Rhône corridor 22
Right/right-wing x, 7–8, 10, 28, 30–1, 34–8, 48, 74, 79–80, 83, 86–90, 92, 95, 98, 115, 127, 129, 134, 148–9, 172–4, 178
 right-wing naturist fundamentalism 79–80
 Right-Left game 172–5
romanticism xiii, 5, 59, 80
Rome 10, 31, 76
Ronsard, Pierre de 3
Rostand, Jean 16
Rougemont, Denis de 12
Rousseau, Jean-Jacques xxxvii, 4–5, 7, 35, 37, 39
 'noble savage' 4, 19, 35
Roy, Christian x, xi n.3, xiv n.6, xviii, xxix, 4 n.2
Ruskin, John 6
Russia 20
Russian Revolution 114, 142

sacred 3, 7, 16, 66, 68, 72, 81, 84, 144, 148

SAFER 132
Sahel 36
Saint-Marc, Philippe 11, 89
Saint-Tropez 189–90
Sartre, Jean-Paul 189 n.24
Sauvage, Le 30
Sauvy, Alfred 26, 181
Savoy 10
Scania 56
school 35, 37–8, 90, 110, 114
Schumacher, E. F 104, 161, 163, 166, 169, 175 see also *Small is Beautiful*
science ix, xiii, xxi, xxiii, xxiv, xxvi, xxviii, xxxv, 10, 16, 18–19, 27, 30, 32–3, 35, 37, 40, 46, 50–1, 61, 72–4, 76, 81, 85, 95, 101–4, 109, 117, 119, 122, 125, 143–5, 152, 157, 160, 181–2, 185
science fiction 6, 84
Scout movement 8
Second Nature xii, xxxiii n.8, 53, 61–3, 85
Second World War 4, 91
sectorialization 114, 131
Seine River 11, 190
self-acceleration xxii
self-management 27 n.12, 36 n.15, 37, 45, 61, 86, 89, 90, 105, 183–5
sex 21, 23, 37, 45, 63, 67, 86–7, 111, 124, 178
sexual revolution 23, 45
Shakers 21
situationism 22–3, 40, 123, 131
Small is Beautiful 104, 161, 163, 166, 169, 175 see also Schumacher, E. F.
SNCF (French railway system) 104
social democracy 97, 128–9
socialism xxiii, 9, 73, 95, 118, 133, 135, 143, 145, 151, 173

INDEX

Socialist Party 35, 134–5
Société française d'écologie 25
sociology/sociologists
 xxxv–xxxvii, 11 n.20, 25 n.8,
 63, 81, 85, 101, 116–17, 172
Sodom 70
soft technology 145, 179 *see also*
 gentle technology
Sogreha 119
solar energy 27, 41, 98, 100,
 104, 119, 120, 153, 179,
 180, 186
Soviet Communist Party 72
Soviet Union 9–10, 110, 189
 see also USSR
Spain 28, 147 n.1, 183 n. 13
SPD (German social
 democracy) 97
spectacle 9, 22–5, 59, 99, 106,
 114–15, 118, 122–3, 131,
 150–1, 154, 165–6, 168
Stalin, Joseph xxxiii, 7, 9, 10, 34,
 38, 116, 117, 125, 129, 137,
 142, 155 n.1
State x, xiii, xxiii, xxv–vi, xxxiii,
 9, 12, 21–2, 26, 34, 37, 48–9,
 53, 61–2, 69, 89–91, 100,
 103–4, 106, 108–11, 119–22,
 124, 130, 132, 135–6, 142–3,
 145, 147–8, 152, 155, 161,
 163, 167, 169, 171, 176–9,
 181–2, 184–5
Stevenson, Robert Louis 5
Stockholm United Nations
 Conference on the Human
 Environment (1972) 20
structuralism xxxv, 19
suburbs 11, 46, 56–7, 107, 186,
 189, 191
surrealism 10, 22, 40, 131
survival of the fittest 67
Survivre et vivre 27, 40
Switzerland 169, 175
synarchy 116

system xv, xxv, xxvi, xxviii, 10,
 62, 83, 86, 88, 92, 98, 101,
 108, 117, 120, 132–3, 141,
 147, 149, 159, 184, 186
systems analysis xxxv, 102

Taylorization 189
technique xvi, 7 n.11, 11, 34, 46,
 61, 72–3, 90, 99–100, 102–4,
 109, 123, 125–6, 128–9, 141,
 143–5, 150 n.3, 163, 180,
 181 n.11, 187
technocracy xiv, 10–11, 24, 26
 n.10, 27, 30, 84, 89, 107–8,
 113, 116–17, 119, 122–4,
 130, 132, 143, 148–9
techno-economic development
 xxv–xxvi
techno-industrial development
 xxiii, xxv–xxvi
technology ix–x, xii–xv,
 xvii–xviii, xxi–xxviii, xxxi, 9,
 37, 74, 100, 145, 179, 185
technoscience x–xi, xiii–xiv, 144–5
technostructure 118
Teilhard de Chardin, Pierre 139
television xxxvii, 13, 25, 103,
 105, 107, 113, 117, 120, 124,
 159, 163–4
terrorism 6, 25 n.7, 90, 99,
 161, 175–6
TGV (*train à grande vitesse*) xxxi,
 27, 100
Third World 11, 20, 26 n.10, 37,
 109, 187
Thoreau, Henry David 5, 16
Tignes dam 10
Torrey Canyon oil spill 26
totalitarianism 22, 30, 62, 83, 86,
 98, 128, 131–2, 137, 141–2,
 144, 182, 186, 189 n.24
total social phenomenon xxii
Touraine (region of France) 56
Touraine, Alain 11, 117

tourism xiii, xxviii, 5, 11, 20–1, 26, 32, 56, 59, 60, 83, 101, 113, 115, 124, 180, 189–91
town 49, 55–7, 161, 169, 188
train xxxi, 27, 100, 104
Trente Glorieuses 26 n.10
Trigano, Gilbert 191
Trotskyites 31, 131
Tuscany 56

uchronia 38
underdevelopment 9, 26 n.10, 36, 126
Union des Démocrates pour la République (UDR) 134
Union pour la démocratie française (UDF) 36
unions 29, 32, 151, 162
United States xiv, xvii, 5–7, 9, 15–16, 18, 20–1, 23, 38–9, 87, 110, 116–17, 119, 128–9
university 118–19
urbanism 89, 101, 122
urbanization 3, 5, 19, 56, 132, 188
USSR 20, 38, 129 *see also* Soviet Union
utopia xxiv, 4
u-topia 38, 86, 141, 143–5, 147–9, 151–3, 162

value creation (*mise en valeur*) 181
Vatican 72
vegetarianism 8, 33, 40, 109
Vendée 180

Verne, Jules 6
Vichy regime 30 n.3, 116 n.5
Vie Claire, La 33
Vietnam 36, 38, 129
villages xxxiii n.8, 10, 38–9, 48–9, 83, 89, 107, 123, 130, 149, 169, 181
violence xxv, 38, 40, 49, 51, 70, 73, 84, 90–1, 110, 114, 132, 151, 162–3, 175
Volga River 20

water-treatment plants 26
Wells, H. G. 6
wilderness xi, xv, xvii, xxvii, 85
women, status of 82
women's liberation 37, 45, 79, 82, 86, 108, 178
Wordsworth, William xiii
work 10–11, 19, 22–3, 34, 48, 56, 58, 67–8, 70, 77, 89, 92, 95, 104–5, 107, 136–7, 154, 156–8, 184, 189–91
workers 10, 27 n.12, 82, 119, 150, 184, 189
World Council of Churches 72
World State xxvi, 143

'Year of the Protection of Nature' 23–4
Year Two Thousand xxxiii n.8, xxxiv, 11, 54, 103, 172, 187

ZAC procedure 31, 61–2
zero growth and zero degrowth 179–80